Lecture Notes in Physics

Volume 965

The Lecture Notes in Physics

The series Lecture Notes in Physics (LNP), founded in 1969, reports new developments in physics research and teaching-quickly and informally, but with a high quality and the explicit aim to summarize and communicate current knowledge in an accessible way. Books published in this series are conceived as bridging material between advanced graduate textbooks and the forefront of research and to serve three purposes:

- to be a compact and modern up-to-date source of reference on a well-defined topic
- to serve as an accessible introduction to the field to postgraduate students and nonspecialist researchers from related areas
- to be a source of advanced teaching material for specialized seminars, courses and schools

Both monographs and multi-author volumes will be considered for publication. Edited volumes should, however, consist of a very limited number of contributions only. Proceedings will not be considered for LNP.

Volumes published in LNP are disseminated both in print and in electronic formats, the electronic archive being available at springerlink.com. The series content is indexed, abstracted and referenced by many abstracting and information services, bibliographic networks, subscription agencies, library networks, and consortia.

Proposals should be sent to a member of the Editorial Board, or directly to the managing editor at Springer:

Dr Lisa Scalone
Springer Nature
Physics Editorial Department
Tiergartenstraße 17
69121 Heidelberg, Germany
lisa.scalone@springernature.com

More information about this series at http://www.springer.com/series/5304

Jonas Larson • Erik Sjöqvist • Patrik Öhberg

Conical Intersections in Physics

An Introduction to Synthetic Gauge Theories

Jonas Larson
Department of Physics
Stockholm University
Stockholm, Sweden

Erik Sjöqvist
Department of Physics and Astronomy
Uppsala University
Uppsala, Sweden

Patrik Öhberg
IPaQS/EPS
Heriot-Watt University
Edinburgh, UK

ISSN 0075-8450   ISSN 1616-6361   (electronic)
Lecture Notes in Physics
ISBN 978-3-030-34881-6   ISBN 978-3-030-34882-3   (eBook)
https://doi.org/10.1007/978-3-030-34882-3

This Springer imprint is published by the registered company Springer Nature Switzerland AG.
The registered company address is: Gewerbestrasse 11, 6330 Cham, Switzerland

To the memory of Stig Stenholm.

Preface

In the history of science we find many examples where the wheel has been reinvented. It may be that some work, for some reason, fell into oblivion or it was simply never recognised by the community. In Einstein's two papers *Quantentheorie des einatomigen idealen Gases* from 1924 and 1925, he essentially predicted condensation. Einstein realised that there is a critical temperature below which a single state becomes macroscopically populated; he calls it 'condensation'. Interestingly, at that time the concept of an order parameter had not been introduced, but Einstein notices 'One can assign a scalar wave field to such a gas'. What Einstein was hinting at was what today is known as the Gross–Pitaevskii equation. It took, however, Gross and Pitaevskii more than 30 years to write down the equation for the classical field. Another example is that of the Aharonov–Bohm effect presented in 1959. However, less known is that the effect was predicted already some 10 years earlier by Ehrenberg and Siday.

The above gives two examples where a result in its full glory has not been recognised (or it is simply not known). This often happens when essentially the same phenomenon is rediscovered in different communities. The 1984 seminal work by Berry on 'phase factors accompanying adiabatic changes' had two precursors. Already in 1956, Pancharatnam demonstrated how the interference of two polarised light beams depends on a geometric phase. While in 1958, Longuet-Higgins et al. showed that adiabatic following around a conical intersection between two electronic molecular potential surfaces resulted in a sign change of the wave function. Berry's work showed the generality of this phenomenon; whenever the wave function adiabatically encircles some sort of singularity in parameter space it acquires a geometric phase factor. Such singularity appears at a conical intersection which is characterised by a point degeneracy of at least two energy surfaces in some parameter space.

Conical intersections are to be found in a range of different physical systems, and it seems that the importance of them has often been analysed independently in the different communities. Thus, the physics of conical intersections can be regarded as yet another example for where the wheel has been reinvented. Even if the concept of conical intersections as such is the same in the different communities, there are still differences between conical intersections in say molecular and in condensed matter physics. In this monograph we gather and discuss various systems where conical intersections have played an important role. The similarities and differences are highlighted, as well as drawing attention to the origin of the physics.

With the ever ongoing experimental progress in the microworld, the boundaries between different fields of physics are becoming ever more diffuse. The cooling and controlled manipulation of individual atoms/ions/molecules/photons have led to an avalanche of new physics, ranging from single particle quantum control to *in situ* explorations of quantum many-body systems. The interaction among hundreds of ultracold atoms or ions can be monitored to such a degree that exotic phases of matter can be experimentally investigated and tailored dynamics of closed quantum many-body systems can for the first time be studied in the lab. It is possible to construct on-chip electric circuits that realise novel photonic lattice models with an effective photon–photon interaction mediated by superconducting quantum dots. This elimination of clear boundaries between the different subfields asks for a better and broader knowledge beyond ones expertise. It is our hope that this book can help in bridging knowledge from different areas in physics.

Two of the main research directions in microscopic physics during the last couple of decades are *quantum simulators* and *topological matter*. Geometric phases and synthetic gauge theories play central roles in both these fields. It turns out that a correct picture of some novel states of matter is only emerging after introducing nonlocal quantities which are directly related to the geometric phase. Many of the proposed quantum simulators consider noncharged particles (like atoms and photons), and in order to simulate, for example, charge particles in a magnetic field one needs to construct synthetic magnetic fields. The physics of conical intersections serves as a very convenient introduction to synthetic gauge theories. The sign change of the wave function upon adiabatically encircling a conical intersection is a manifestation of a geometric phase, and it may be envisioned as if a magnetic flux is penetrating the point of the conical intersection. In this sense, it bares many similarities to the Aharonov–Bohm effect.

This book has been written with the hope that the reader must not be an expert in some particular field in order to grasp its content. Ideally, it should be possible to understand the ideas in say molecular or condensed matter physics without having been particularly exposed to those fields. Basic knowledge in quantum mechanics is, however, a prerequisite. Chapter 2 presents the general background to the topic. Central is the concept of adiabaticity, which is essential in order to fully appreciate the following chapters. However, it is not crucial to know Born–Oppenheimer theory to understand the content of Chap. 4 on condensed matter physics. The following four chapters can be read individually, even if there exist a few cross-references between them in order to tie them together.

We have focused on three main fields in physics where conical intersections are found. In Chap. 3 we discuss the basics of conical intersections in molecular physics. It was also in this field that their importance was first discussed. They mark the breakdown of the Born–Oppenheimer approximation and give rise to many observable effects. The following Chap. 4 is devoted to conical intersections in condensed matter physics. A crucial difference between conical intersections in molecular and condensed matter physics is that in the former they appear in real space while in the latter they are found in (quasi) momentum space. The example known for most people is probably that of Dirac cones in graphene, but we also discuss how they come about in spin–orbit coupled systems and how they generalise to higher dimensions, so-called Weyl points. In Chap. 5, conical intersections in systems of cold atoms are analyzed. Like for molecules, the underlying theory here is that of Born–Oppenheimer. But in comparison to molecules, in these systems there is a greater freedom to manipulate the actual model Hamiltonians which in certain cases allows for more clean experimental tests.

In the book we also include a chapter on other physical systems where conical intersections can emerge. In particular, it is shown how Jahn–Teller models that appear in molecular physics are greatly related to the Jaynes–Cummings model which forms a backbone of cavity quantum electrodynamics and trapped ion physics. One section of this chapter is assigned to open quantum systems. Only within the last years we have seen an enormous interest in 'non-Hermitian' quantum mechanics and the importance of *exceptional points*. Such exceptional points can be seen as conical intersections in the complex plane, and they typically show up in the theory for open quantum systems.

Acknowledgements

Over the years, there have been many people who we have collaborated or discussed with, people whom, in one way or another, have had an influence on this book. We are especially thankful to Alexander Altland, Emil Bergholtz, Jean Dalibard, Marie Ericsson, Barry Garraway, Gonzalo García de Polavieja, Nathan Goldman, Osvaldo Goscinski, Hans Hansson, Niklas Johansson, Gediminas Juzeliūnas, Thomas Klein Kvorning, Åsa Larson, Maciej Lewenstein, Jani-Petri Martikainen, Luis Santos, Ian Spielman, Stig Stenholm, Robert Thomson, Dianmin Tong, Manuel Valiente, and Johan Åberg.

Stockholm, Sweden   Jonas Larson
Uppsala, Sweden   Erik Sjöqvist
Edinburgh, UK   Patrik Öhberg
September 2019

Contents

Chapter 1
Introduction

A rather old-fashion view of quantum mechanics is that it is an application of the physics of waves. A not so old-fashion way to get interesting physics out of mathematical objects is by studying their singularities. [1]

Conical intersections (CIs) are singularities of the energy spectra of parameter-dependent quantum systems. The term refers to the formation of conically shaped energy surfaces in the vicinity of points in parameter space where two or more energies cross. CIs are important in physical settings characterised by slow and fast degrees of freedom, such as the fast electrons and slow nuclei of a molecule, but also for the evolution of quantum systems controlled by slow classical parameters, as well as for the understanding topological characteristics of condensed matter systems. In this way, the importance of CIs covers a significant portion of modern quantum physics.

The CI concept appeared implicitly already in the late 1930s in the work by Jahn and Teller on the stability of symmetric molecular configurations [2]. Longuet-Higgins and coworkers examined the dynamical consequences of CIs, such as a fractional quantisation of the coupled electron-nuclear (vibronic) motion in polyatomic molecules [3]. The analogy between the phase behaviour of an adiabatic wave function in the vicinity of a CI and the Aharonov–Bohm effect was described by Mead in the early 1980s [4]. But it was not until Berry demonstrated the generality of the geometric phase and the gauge structure underlying adiabatic time-evolution, that the importance of CIs was fully realised [5]. This insight has led to several important new findings, such as new topological states of matter [6], gauge structures governing the dynamics of cold neutral atoms in various trapping geometries [7], and new forms of robust quantum computation [8].

In these Lecture Notes, we discuss the theory of CIs and the associated synthetic gauge theories with applications to a range of physical systems. We focus on essentially three different contexts where CIs are known to play a major role: molecular physics, condensed matter systems, and trapped cold atoms. A key

© Springer Nature Switzerland AG 2020
J. Larson et al., *Conical Intersections in Physics*, Lecture Notes in Physics 965,
https://doi.org/10.1007/978-3-030-34882-3_1

objective of the Lecture Notes is to emphasis the unifying aspect of CIs and synthetic gauge structures in quantum physics.

Chapter 2 is devoted to the concept of adiabaticity in quantum systems. In Sect. 2.2, we discuss the theory of adiabatic time-evolution and examine some of its subtleties. The synthetic gauge structure associated with parameter-dependent energy subspaces is described in Sect. 2.3. This gauge structure is generally non-Abelian, meaning that it gives rise to matrix-valued 'phase factors' (more precisely, unitary matrices), but becomes Abelian in the case of non-degenerate energies. Here, we see the first example where CIs play a central role, namely that these intersections are sources of magnetic monopoles, which are missing entities in electromagnetism, but now can be studied experimentally in adiabatic systems.

In Sect. 2.4, we discuss the time-independent counterpart of adiabatic systems. Here, the parameters are themselves quantum variables, such as the nuclear degrees of freedom in a molecule or crystal. In this setting, the slow coordinates move on adiabatic potential energy surfaces (APSs), being parameter dependent solutions of the time-independent Schrödinger equation for the fast variables. Such 'fast–slow' quantum systems are called Born–Oppenheimer systems. Just as in the time-dependent case, the Born–Oppenheimer setting gives rise to synthetic gauge fields, but now having a direct physical effect on the motion of the slow system.

In Chap. 3, we further explore the synthetic gauge structure in Born–Oppenheimer systems, by considering the molecular case in more detail. As mentioned above, the nuclear variables in a molecule play the role of slow parameters coupled to the fast electronic degrees of freedom. In Sect. 3.2.1, we examine different types of intersections between electronic APSs and in Sect. 3.2.2, we discuss tests to find intersection points by using tools from algebraic topology.

Sections 3.3 and 3.4 focus on the Jahn–Teller effect and its relation to CIs between symmetry adapted electronic states. We discuss the linear+quadratic $E \times \epsilon$ Jahn–Teller system, which, despite its simplicity, is associated with a quite intricate CI structure. We examine the Berry phase as well as the dynamical implications of the CI structure for the nuclear pseudo-rotational motion. In particular, we demonstrate that this system provides a realisation of a spin Hall effect that is manifest in the dynamics of the nuclei.

In Chap. 4, we turn our attention to condensed matter systems. An essential conceptual difference between the CIs considered up to this point and those in condensed matter systems is that the latter appear in momentum space. In other words, a CI is a point or line in the Brillouin zone where two or more energy bands cross. The perhaps most well-known example of such a crossing is for graphene, i.e., a hexagonal monolayer of carbon atoms for which the Fermi surface is exactly at the CIs. By linearising the dispersion, the electrons close to the Fermi level are described by a Dirac-like equation in $2 + 1$ dimensions, but with an effective 'speed of light' being considerably smaller than c. CIs in the context of condensed matter systems are therefore called Dirac points or Dirac CIs.

Dirac CIs are also of importance in spin–orbit coupled systems and in the BCS theory of superconductivity. Another central concept, used in order to analyse condensed matter systems with non-trivial topological features, is the Chern number. This quantity is a topological invariant of an energy band in two dimensions and is defined as the Berry phase integrated over the whole Brillouin zone. Its importance is evident as it is directly related to the quantised Hall conductivity.

The motion of cold atoms in slowly varying inhomogeneous laser fields is governed by artificial gauge fields that arise when the lasers induce fast transitions between the internal atomic levels. These gauge fields can be Abelian or non-Abelian. In Chap. 5, we discuss the slow motion of cold atoms in inhomogeneous optical fields and some of the peculiar effects associated with the artificial gauge structure. Standing laser wave fields can be used to create optical lattice that can be used to simulate real crystals by injecting atoms into the periodic structure and thereafter studying their motion.

Sections 5.2 and 5.3 contain a description of light–atom interaction and the adiabatic dynamics associated with the induced synthetic gauge fields. The idea is to view the atom as a Born–Oppenheimer-like system for which the centre of mass and the internal energy levels play the role as the slow and fast degrees of freedom, respectively. These 'subsystems' are coupled via the inhomogeneity of the laser fields, which are assumed to vary sufficiently slowly for the system to be adiabatic. Just as in the spin systems discussed in Chap. 2, the induced gauge fields take the form of magnetic monopoles located at CIs, now in the space defined by the laser parameters. We discuss, in Sect. 5.4, how non-Abelian synthetic gauge fields can be obtained if the lasers couple three ground state levels to an excited stated, forming a tripod system. These gauge fields can be used to simulate relativistic effects, such as Zitterbewegung, which otherwise require relativistic particles.

The synthetic gauge fields are not limited to single atom systems. In Sect. 5.5, we discuss the theory of Bose–Einstein condensates, in which a cloud of atoms undergo a phase transition so as to behave as a macroscopically occupied single quantum state at sufficiently low temperatures. The description of a condensate differs significantly from fundamental quantum systems in that the atom–atom collisions give rise to non-linear effects, as described by the Gross–Pitaevskii equation. The non-linearity implies that CIs can have completely different shapes in parameter space, as discussed in Sect. 5.5.2.

In Chap. 6, we examine CIs in some other system. We first consider cavity electrodynamics in Sect. 6.1. Here, the internal levels of an atom are coupled to a quantised cavity field. This system is described by the Rabi or Jaynes–Cummings models. In the semi-classical regime, one obtains CIs in the quadrature phase space. In particular, we demonstrate how such a system can give rise to an intrinsic anomalous Hall effect.

In Sect. 6.2, we discuss a system of trapped ions interacting with the vibrations in the trap. The ion trap and cavity settings both describe a discrete level system interacting with a harmonic oscillator system, in the former case the oscillations in the trap while in the latter case the oscillations of the photon standing wave in the cavity. We show how the ion trap system can be used to realise different types of Jahn–Teller systems.

The relevance of CIs is not restricted to quantum systems. They may occur also in classical wave systems. In Sect. 6.3, we discuss how such intersections may show up for classical light of sufficiently long wavelength moving through materials with a small space-dependent refractive index and how lattice models can be emulated in such settings. These systems may realise Dirac cones by appropriately choosing the lattice structure. Among other conical singularity structures in classical systems, but not covered by these Lecture Notes, are found in Einstein's theory of gravity. These structures occur along the world sheet of cosmic strings [9] and for black holes [10]. They differ conceptually from those considered here in the sense that they contain a single cone and not two intersecting ones.

In Sect. 6.4, we end our expose over CIs in quantum physics by considering what happens if the system is open. An open system is defined as a quantum system that interacts with some kind of 'environment'. This means that we need to replace the Schrödinger equation by an effective description, which we obtain by tracing over the environmental degrees of freedom. Under certain approximations (essentially the Markov limit), this is described by the Lindblad equation, which is a master equation that models the evolution of the density operator. The Liouvillian of a Lindblad equation is not Hermitian, and the energies typically become complex-valued. Thereby, the concept of CIs drastically change. Still there is a notion of energy level crossings in parameter space, called exceptional points, which are essentially CIs where both the real and imaginary parts of the spectrum cross simultaneously.

References

1. Berry, M.V.: Wave geometry: a plurality of singularities. In: Anandan, J.S. (ed.) Quantum Coherence, pp. 92–98. World Scientific, Singapore (1991)
2. Jahn, H.A., Teller, E.: Stability of polyatomic molecules in degenerate electronic states. I. Orbital degeneracy. Proc. R. Soc. A. **161**, 220 (1937)
3. Longuet-Higgins, H.C., Öpik, U., Pryce, M.H.L., Sack, R.A.: Studies of the Jahn–Teller effect. II. The dynamical problem. Proc. R. Soc. Lond. Ser. A **244**, 1 (1958)
4. Mead, C.A.: The molecular Aharonov-Bohm effect in bound states. Chem. Phys. **49**, 23 (1980)
5. Berry, M.V.: Quantal phase factors accompanying adiabatic evolution. Proc. R. Soc. Lond. Ser. A **392**, 45 (1984)
6. Bernevig, B.A.: Topological Insulators and Topological Superconductors. Princeton University Press, Princeton and Oxford (2013)
7. Dalibard, J., Gerbier, F., Juzeliūnas, G., Öhberg, P.: *Colloquium*: artificial gauge potentials for neutral atoms. Rev. Mod. Phys. **83**, 1523 (2011)
8. Carollo, A.C.M., Vedral, V.: Holonomic quantum computation. In: Bruß, D., Leuchs, G. (eds.) Quantum Information: From Foundations to Quantum Technology Applications, pp. 475–482. Wiley-VCH Verlag GmbH & Co. KGaA (2019)
9. Oliveira-Neto, G.: Identifying conical singularities. J. Math. Phys. **37**, 4716 (1996)
10. Solodukhin, S.N.: Conical singularity and quantum corrections to the entropy of a black hole. Phys. Rev. D **51**, 609 (1995)

Chapter 2
Theory of Adiabatic Evolution

Abstract We identify two scenarios where the concept of adiabaticity emerges, differing on whether the slow parameters are themselves dynamical variables or not. In both cases, the separation of time-scales leads to synthetic gauge structures that have measurable consequences. We describe the basic theory of adiabatic evolution. Conical intersections, which are points in the space of slow parameters where two or more energies cross, are of central importance for the understanding of the emergent synthetic gauge fields in adiabatic systems.

2.1 Introduction

Solving the Schrödinger equation is a central problem in physics, since it determines the allowed energies and the time-evolution of quantum-mechanical systems. However, to find these solutions is a difficult task if the involved Hamiltonian contains many degrees of freedom or has a non-trivial time-dependence. These features make the problem to find exact solutions of the Schrödinger equation quickly intractable in practise. Thus, simplifying approximations are highly needed. Our objective here is to describe the physics related to adiabatic evolution, which is an approximate form of dynamics of quantum-mechanical systems that has applications in essentially all subfields of quantum physics.

Adiabaticity basically quantifies the notion of *slowness* and applies to situations where different quantities vary on vastly different time-scales. It appears in two classes of situations in quantum physics:

- Time-dependent quantum-mechanical systems that are driven by slowly changing classical parameters (for instance, the phase and amplitude of a laser field that drives transitions in an atom).
- Time-independent composite quantum-mechanical systems that can be divided into subsystems associated with very different masses (for instance, the electrons and nuclei of a molecule or a crystal).

Slowness in the latter of these situations is related to the fact that objects with 'large' mass, in a relative sense, typically move slower than objects with 'small' mass.

© Springer Nature Switzerland AG 2020

J. Larson et al., *Conical Intersections in Physics*, Lecture Notes in Physics 965,
https://doi.org/10.1007/978-3-030-34882-3_2

We shall see that the notion of adiabaticity is associated with rich physical properties and conceptual subtleties. In this chapter, we outline the basic theory for adiabatic evolution and provide some examples. We start with time-dependent systems in Sect. 2.2, where we delineate the conditions under which adiabatic evolution occurs. Section 2.3 shows how a *synthetic* gauge structure, similar to that of electromagnetism, appears in the adiabatic regime of time-dependent systems. Section 2.4 addresses the time-independent situation, leading to the *Born–Oppenheimer approximation*. It turns out that this approximation can be associated with a similar kind of synthetic gauge theory as in the time-dependent case. In this way, the synthetic gauge structure serves as a unifying aspect of adiabatic systems.

2.2 Adiabatic Time-Evolution

Adiabatic time-evolution of a quantum-mechanical object is enforced by slowly varying some experimental control parameters $\mathbf{R}$ (such as the phases and amplitudes of a set of laser beams, or the direction of a magnetic field) along a path $\mathscr{C} : [0, 1] \ni s \mapsto \mathbf{R}(s)$ in parameter space. Such a process is described by a continuous family of Hamiltonians $\hat{H}(\mathbf{R}(t/T))$, $t \in [0, T]$. Here, the run-time T roughly measures the 'slowness' of the parameter change: if T is much larger than the minimal energy gap of the Hamiltonian family, then the parameters are said to undergo a 'slow' or 'adiabatic' change.

2.2.1 Adiabatic Theorem

The precise meaning of 'slowness' is given by the adiabatic theorem. Here, we state this important theorem and specify under which premises it holds.

The adiabatic theorem gives the rate at which transitions between the eigenspaces of $\hat{H}(\mathbf{R}(t/T))$ tend to zero when $T \rightarrow \infty$. Explicitly, by means of the parameter change $t \mapsto s = t/T$, the adiabatic theorem can be stated as

$$\hat{U}_T(s, 0) \hat{P}_n(\mathbf{R}(0)) - \hat{P}_n(\mathbf{R}(s)) \hat{U}_T(s, 0) = O(1/T), \quad T \rightarrow \infty, \ \forall n, \quad (2.1)$$

where $\hat{P}_n(\mathbf{R}(s))$ is an eigenprojector of $\hat{H}(\mathbf{R}(s))$ associated with the energy eigenvalue $\varepsilon_n(s)$ and

$$\hat{U}_T(s, 0) = \mathscr{T} e^{-iT \int_0^s \hat{H}(\mathbf{R}(s')) ds'} \quad (2.2)$$

is the exact time-evolution operator ($\mathscr{T}$ denotes time-ordering; we put $\hbar = 1$ throughout this chapter). The theorem can be proved (see, for instance, [1]) provided the following three premises holds:

(i) The eigenvalues $\varepsilon_n(\mathbf{R}(s))$ of the instantaneous Hamiltonian are continuous functions of s.
(ii) The eigenvalues do not cross on $s \in [0, 1]$.
(iii) $\frac{d}{ds}\hat{P}_n(\mathbf{R}(s))$ and $\frac{d^2}{ds^2}\hat{P}_n(\mathbf{R}(s))$ are well-defined and piecewise continuous.

An implication of these conditions is that the rank of $\hat{P}_n(\mathbf{R}(s))$ is not allowed to change, not even at isolated points, on $s \in [0, 1]$.

To see what the adiabatic theorem means, let us consider the simplest case where the eigenprojector is one-dimensional (rank = 1) so that we can write $\hat{P}_n(\mathbf{R}(s)) = |\psi_n(\mathbf{R}(s))\rangle\langle\psi_n(\mathbf{R}(s))|$ throughout the evolution. If the system starts in the state $|\Phi(0)\rangle = |\psi_n(\mathbf{R}(0))\rangle$ and all the premises (i)–(iii) hold, then (2.1) implies

$$|\Phi(\mathbf{R}(s))\rangle = \hat{U}_T(s, 0)|\psi_n(\mathbf{R}(0))\rangle$$

$$= |\psi_n(\mathbf{R}(s))\rangle\langle\psi_n(\mathbf{R}(s))|\hat{U}_T(s, 0)|\psi_n(\mathbf{R}(0))\rangle + O(1/T) \quad (2.3)$$

when $T \to \infty$. We thus see that if the system starts in an energy eigenstate, it remains in an energy eigenstate for $s \in [0, 1]$. We further note that $U_T(s, 0)$ preserves the norm of the state, which implies

$$\langle\psi_n(\mathbf{R}(s))|\hat{U}_T(s, 0)|\psi_n(\mathbf{R}(0))\rangle \to e^{i\kappa_n(s)} \quad (2.4)$$

in the adiabatic $T \to \infty$ limit. We shall have more to say about the phase $\kappa_n(s)$ in Sect. 2.3.3 below.

2.2.2 Adiabatic Approximation

The adiabatic theorem gives the rate at which a given state starting in an eigensubspace of the system Hamiltonian remains in the eigensubspace when the Hamiltonian change becomes increasingly slow. However, it should be clear that the state never follows the instantaneous energy eigenstates perfectly for a large but finite T; there are always small but non-zero non-adiabatic transitions between the eigensubspaces. Thus, perfect adiabatic evolution is an idealisation and in practical applications it becomes essential to tell how well the evolution coincides with the adiabatic one. Here, we find a useful condition that answers this question.

Consider a quantum system with Hilbert space $\mathscr{H}$ and parameter-dependent Hamiltonian $\hat{H}(\mathbf{R}(s))$. Let $\hat{P}_n(\mathbf{R}(s)) = \sum_{k=1}^{g_n}|\psi_{n;k}(\mathbf{R}(s))\rangle\langle\psi_{n;k}(\mathbf{R}(s))|$, $n = 1, \ldots, K \le \dim\mathscr{H}$, be the eigenprojectors of $\hat{H}(\mathbf{R}(s))$, g_n being the rank (degree of degeneracy) of the nth eigensubspace, with corresponding energies $\varepsilon_n((\mathbf{R}(s)))$. The eigensubspaces span the Hilbert space, meaning that any solution $|\Phi(s)\rangle$ of the reparametrised ($t \mapsto s = t/T$) Schrödinger equation

$$i|\dot{\Phi}(s)\rangle = T\hat{H}(\mathbf{R}(s))|\Phi(s)\rangle \tag{2.5}$$

can be expanded in the energy eigenstates, i.e. we may write

$$|\Phi(s)\rangle = \sum_{n=1}^{K}\sum_{k=1}^{g_n} e^{-iT\int_0^s \varepsilon_n(s')ds'}|\psi_{n;k}(\mathbf{R}(s))\rangle c_{n;k}(s). \tag{2.6}$$

By inserting this into (2.5) and integrating, we obtain

$$c_{n;k}(s) = c_{n;k}(0) - \sum_{l=1}^{g_n}\int_0^s \langle\psi_{n,k}(\mathbf{R}(s'))|\dot{\psi}_{n,l}(\mathbf{R}(s'))\rangle c_{n;l}(s')ds'$$

$$- \sum_{m\neq n=1}^{K}\sum_{l=1}^{g_n}\int_0^s e^{iT\int_0^{s'}\Delta_{nm}(\mathbf{R}(s''))ds''}\langle\psi_{n,k}(\mathbf{R}(s'))|\dot{\psi}_{m,l}(\mathbf{R}(s'))\rangle c_{m;l}(s')ds' \tag{2.7}$$

with the energy gap functions $\Delta_{nm} = \varepsilon_n - \varepsilon_m$. The third term on the right-hand side of (2.7) describes non-adiabatic transitions between the energy eigensubspaces and should be negligible in the adiabatic regime. It contains two competing time-scales: the rate of transition between the eigensubspaces

$$\frac{1}{\delta s}\left|\langle\psi_{n,k}(\mathbf{R}(s))|\psi_{m,l}(\mathbf{R}(s+\delta s))\rangle\right| = \left|\langle\psi_{n,k}(\mathbf{R}(s))|\dot{\psi}_{m,l}(\mathbf{R}(s))\rangle\right| \tag{2.8}$$

and the 'intrinsic' oscillation frequency for transitions between the eigensubspaces of the instantaneous Hamiltonian

$$\left|T\frac{d}{ds}\int_0^s ds'\Delta_{nm}(\mathbf{R}(s'))\right| = T|\Delta_{nm}(\mathbf{R}(s))| \tag{2.9}$$

associated with the phase factors $e^{iT\int_0^{s'}\Delta_{nm}(\mathbf{R}(s''))ds''}$. Intuitively, the non-adiabatic term of (2.7) washes out due to these oscillatory phase factors when the intrinsic frequency is much larger than the transition rate. This yields the adiabatic condition,

$$T \gg \left|\frac{\langle\psi_{n,k}(\mathbf{R}(s))|\dot{\psi}_{m,l}(\mathbf{R}(s))\rangle}{\Delta_{nm}(\mathbf{R}(s))}\right|, \quad \forall s \in [0,1], \; \forall m. \tag{2.10}$$

By using the eigenvalue equation $\hat{H}(\mathbf{R}(s))|\psi_{n;k}(\mathbf{R}(s))\rangle = \varepsilon_n(\mathbf{R}(s))|\psi_{n;k}(\mathbf{R}(s))\rangle$, one can write the adiabatic condition on the alternative form

$$T \gg \left|\frac{\langle\psi_{n,k}(\mathbf{R}(s))|\frac{d}{ds}\hat{H}(\mathbf{R}(s))|\psi_{m,l}(\mathbf{R}(s))\rangle}{\Delta_{nm}^2(\mathbf{R}(s))}\right|, \quad \forall s \in [0,1], \; \forall m. \tag{2.11}$$

The adiabatic condition is necessary to guarantee the validity of the adiabatic approximation [2]. This approximation entails that for large but finite T any initial state that starts in an eigensubspace approximately follows this eigenstate throughout the evolution. In other words,

$$|\Phi(0)\rangle = |\psi_{n;k}(\mathbf{R}(0))\rangle \mapsto |\Phi(s)\rangle \approx \sum_{l=1}^{g_n} |\psi_{n;l}(\mathbf{R}(s))\rangle c_{n;l}(s)$$

$$\Rightarrow T \gg \left| \frac{\langle \psi_{n,k}(\mathbf{R}(s))|\dot{\psi}_{m,l}(\mathbf{R}(s))\rangle}{\Delta_{nm}(\mathbf{R}(s))} \right|, \quad \forall s \in [0, 1]. \tag{2.12}$$

However, it is important to note that the reverse is not true since the relation between the adiabatic condition and the adiabatic approximation is valid under one important caveat: the adiabatic theorem should hold in the $T \to \infty$ limit, which means, in particular, that all the premises (i)–(iii) (see Sect. 2.2.1) must be satisfied by the Hamiltonian of the system. The Marzlin–Sanders paradox [3], as will be discussed in the following section, illustrates this point in a striking way.

2.2.3 The Marzlin–Sanders Paradox

In an instructive attempt to problematize the notion of adiabaticity in quantum mechanics, Marzlin and Sanders [3] constructed a model system that satisfies the adiabatic condition but fails to generate adiabatic evolution. As we shall see, their construction demonstrates a subtle consequence of when some of the premises of the adiabatic theorem fails.

The starting point of the Marzlin–Sanders argument is a one-parameter Hamiltonian family $\hat{H}(s)$, $s \in [0, 1]$, that drives the evolution of a quantum-mechanical system (any possible $\mathbf{R}$-dependence in the system is irrelevant for the argument and is therefore omitted for notational simplicity). This Hamiltonian family is applied for a run-time T. We assume the corresponding energies $\varepsilon_n(s)$ and energy eigenvectors $|\psi_{n;k}(s)\rangle$ satisfy the premises of the adiabatic theorem; thus, for T satisfying the adiabatic condition in (2.10), the system undergoes approximate adiabatic evolution.

Next, define the 'mirror' Hamiltonian

$$\tilde{\hat{H}}(s, T) = -\hat{U}_T^{\dagger}(s, 0)\hat{H}(s)\hat{U}_T(s, 0), \tag{2.13}$$

$\hat{U}_T(s, 0)$ being the exact time-evolution operator associated with $\hat{H}(s)$. Note that, unlike $\hat{H}$, $\tilde{\hat{H}}$ does not define a fixed one-parameter family of Hamiltonians for all T, i.e. $\tilde{\hat{H}}$ is a function of both s and T. Conceptually, this means that the notion of 'slowness' becomes ambiguous; indeed, one can see from its definition that $\tilde{\hat{H}}(s, T)$

oscillates in an increasingly rapid fashion when T grows. Nevertheless, the energy gaps $\widetilde{\Delta}_{nm}(s) = -\Delta_{nm}(s)$ and energy eigenstates $|\widetilde{\psi}_{n;k}(s)\rangle = \hat{U}_T^\dagger(s, 0)|\psi_{n;k}(s)\rangle$ of the mirror system satisfy the same adiabatic condition as in the original system,

$$
\left| \frac{\langle \widetilde{\psi}_{n;k}(s)|\dot{\widetilde{\psi}}_{m;l}(s)\rangle}{\widetilde{\Delta}_{nm}(s)} \right|
$$
$$
= \left| \frac{i T \langle \psi_{n;k}(s)|\hat{H}(s)|\psi_{m;l}(s)\rangle + \langle \psi_{n;k}(s)|\dot{\psi}_{m;l}(s)\rangle}{-\Delta_{nm}(s)} \right|
$$
$$
= \left| \frac{\langle \psi_{n;k}(s)|\dot{\psi}_{m;l}(s)\rangle}{\Delta_{nm}(s)} \right| \ll T, \tag{2.14}
$$

where we have used the Schrödinger equation $i\hat{U}_T(s, 0)\frac{\partial}{\partial s}\hat{U}_T^\dagger(s, 0) = \hat{H}(s)$ of the original system.

Now, we ask: does the mirror system perform approximate adiabatic evolution for run-time T satisfying (2.14)? In other words, does the adiabatic condition *imply* the adiabatic approximation in this mirror system? To see whether this is the case, we note that adiabatic evolution means that transitions between the energy subspaces are negligible. For the original system, this is equivalent to

$$
\left| \langle \psi_{m\neq n;l}(s)|\hat{U}_T(s, 0)|\psi_{n;k}(0)\rangle \right| \approx 0, \; \forall s \in [0, 1], \; \forall k, l. \tag{2.15}
$$

Let us test whether this holds for the mirror system. We note that $\widetilde{\hat{U}}_T(s, 0) = \hat{U}_T^\dagger(s, 0)$, which implies

$$
\left| \langle \widetilde{\psi}_{m\neq n;l}(s)|\widetilde{\hat{U}}_T(s, 0)|\widetilde{\psi}_{n;k}(0)\rangle \right| = \left| \langle \psi_{m\neq n;l}(s)|\hat{U}_T(s, 0)\hat{U}_T^\dagger(s, 0)|\psi_{n;k}(0)\rangle \right|
$$
$$
= \left| \langle \psi_{m\neq n;l}(s)|\psi_{n;k}(0)\rangle \right|, \tag{2.16}
$$

which can take any value between 0 and 1. We thus arrive at the seemingly paradoxical conclusion that although the mirror system satisfies the adiabatic condition, it does not, in general, evolve adiabatically. That is, the adiabatic condition does not imply the validity of the adiabatic approximation.

The origin of the paradox can be seen by noting that not all premises (i)–(iii) of the adiabatic theorem (see Sect. 2.2.1) are satisfied by the mirror Hamiltonian [4]. Specifically, premises (i) and (ii) concern the energy gaps and are therefore equally valid for $\widetilde{\hat{H}}(s, T)$ and $\hat{H}(s)$ since $\widetilde{\varepsilon}_n(s) = -\varepsilon_n(s)$. The critical premise is instead (iii), which holds, by assumption, for $H(s)$ but not for $\widetilde{\hat{H}}(s, T)$. The reason is that the eigenprojectors $\widetilde{\hat{P}}_n(s, T) = \hat{U}_T^\dagger(s, 0)\hat{P}_n(s)\hat{U}_T(s, 0)$ of the mirror Hamiltonian contain exponential factors, originating from $\hat{U}_T(s, 0)$, whose argument tends to infinity when $T \to \infty$. These arguments make the derivatives of $\widetilde{\hat{P}}_n(s, T)$ undefined in the adiabatic limit, thus invalidating the prerequisites for the adiabatic theorem. One can therefore not expect that the adiabatic condition to have any relation to the adiabatic approximation for this model system.

2.2.4 The Importance of the Energy Gap: Local Adiabatic Quantum Search

The energy gaps $\Delta_{nm}(s)$ are essential for the adiabatic approximation. These gaps should be continuous functions, should never vanish, and should be sufficiently large so as to prevent transitions between different energy subspaces during the evolution. This latter aspect can be used for performing computation. As an illustration of the role of the gap for adiabaticity, we shall consider one such computational problem, which is how adiabatic evolution can be used to search for a marked item in an unstructured database in an efficient way.

Suppose we wish to find a single marked item v in an unstructured list of N entries. This problem can be mapped to a fixed orthonormal basis $\{|k\rangle\}_{k=1}^{N}$ spanning an N-dimensional Hilbert space of a quantum system. We consider the one-parameter family of Hamiltonians

$$\hat{H}(s) = -(1-s)|\Psi\rangle\langle\Psi| - s|v\rangle\langle v|, \quad |\Psi\rangle = \frac{1}{\sqrt{N}}\sum_{k=1}^{N}|k\rangle, \qquad (2.17)$$

where $s = t/T \in [0, 1]$, T being the run-time. $\hat{H}(s)$ has a decoupled $(N-2)$-fold degenerate zero energy excited eigensubspace and two lower non-degenerate eigenstates $|E_g(s)\rangle$ and $|E_e(s)\rangle$ with corresponding energies $E_g(s) < E_e(s) < 0$, defining the non-zero energy gap function,

$$\Delta(s) = E_e(s) - E_g(s) = \sqrt{\frac{1 + (N-1)(2s-1)^2}{N}} \geq \frac{1}{\sqrt{N}}, \qquad (2.18)$$

the lower bound reached at $s = \frac{1}{2}$, thus $\Delta_{\min} = \Delta(\frac{1}{2}) = \frac{1}{\sqrt{N}}$. By preparing the system at $s = 0$ in the ground state $|\Phi(0)\rangle = |E_g(0)\rangle = |\Psi\rangle$, the state ends up with high probability in the ground state of $\hat{H}(1)$ if the evolution is performed adiabatically. Since this ground state is $|E_g(1)\rangle = |v\rangle$, we can find the marked item by measuring in the $|k\rangle$ basis at $s = 1$. More precisely, if we wish to find the marked state $|v\rangle$ with a chosen accuracy $\epsilon \ll 1$, meaning that the success probability is

$$|\langle v|\Phi(1)\rangle|^2 \geq 1 - \epsilon^2, \qquad (2.19)$$

then T must be chosen so that

$$\frac{\max \left|\langle E_e(s)|\frac{d}{ds}\hat{H}(s)|E_g(s)\rangle\right|}{T\Delta_{min}^2(s)} = \frac{\max \left|\langle E_e(s)|\frac{d}{ds}\hat{H}(s)|E_g(s)\rangle\right|}{TN^{-1}} \leq \epsilon \ll 1, \quad (2.20)$$

where we have used (2.18). By diagonalising $\hat{H}(s)$, one can show that

$$\left|\langle E_e(s)|\frac{d}{ds}\hat{H}(s)|E_g(s)\rangle\right| \leq 1, \quad (2.21)$$

which combined with (2.20) implies

$$T \geq \frac{N}{\epsilon}. \quad (2.22)$$

Thus, the time required to find the marked item with a given accuracy ϵ scales as N, which coincides with the scaling of classical search.

However, we can do better by optimising the adiabatic evolution. The idea is to vary the speed so that the evolution slows down when the minimal gap between the ground and first excited states is approached. In other words, one looks for time-local reparametrisations $t \to s = s(t) \in [0, 1]$ that can improve how the run-time T scales with the size N of the database. This is achieved by requiring the adiabatic condition in (2.20) to hold *locally* at each infinitesimal time step, i.e.

$$|\dot{s}(t)| \frac{\left|\langle E_e(s)|\frac{d}{ds}\hat{H}(s)|E_g(s)\rangle\right|}{\Delta^2(s)} \leq \epsilon. \quad (2.23)$$

Equation (2.21) implies that $\dot{s}(t) = \epsilon \Delta^2(s)$ satisfies (2.23) and can be integrated to yield

$$t(s) = \frac{1}{2\epsilon}\frac{N}{\sqrt{N-1}}\left(\arctan(\sqrt{N-1}(2s-1)) + \arctan\sqrt{N-1}\right). \quad (2.24)$$

It can be proved that (2.24) is the optimal choice of reparametrisation [5]. We show the function $s(t)$ on $[0, T]$ in Fig. 2.1 for $N = 4, 16, 36$, and 64 with 96% success probability (i.e. $\epsilon = 0.2$). The run-time of the search is

$$T = t(1) = \frac{1}{\epsilon}\frac{N}{\sqrt{N-1}}\arctan\sqrt{N-1} \geq \frac{\pi}{2\epsilon}\sqrt{N} \quad (2.25)$$

for large N. The linear scaling with $\sqrt{N}$ is clearly visible in Fig. 2.1. This quadratic speed-up is achieved by utilising the central role of energy gaps in adiabatic evolution.

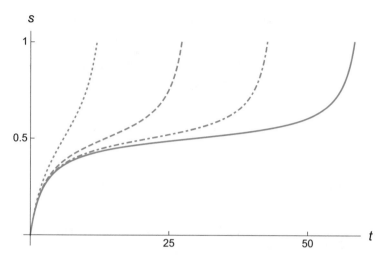

Fig. 2.1 Scaling of the run-time for searching a marked item out of N items, where $N = 4$ (dotted), $N = 16$ (dashed), $N = 36$ (dashed-dotted), and $N = 64$ (solid). The accuracy is set to $\epsilon = 0.2$, meaning that the success probability is 96%. We see that the run-time $T = t(1)$ scales linearly with $\sqrt{N}$. Note how the system spends increasingly more time in the vicinity of the minimum energy gap at $s = \frac{1}{2}$ in order to satisfy the local adiabatic condition in (2.23)

2.3   Gauge Structure of Time-Dependent Adiabatic Systems

A synthetic gauge structure, similar to that of electromagnetism, emerges in adiabatic time-evolution. This gauge structure is associated with the curved geometry of the quantum-mechanical state space and plays a crucial role in the understanding of adiabatic change. For degenerate energy eigensubspaces, as first demonstrated by Wilczek and Zee [6], the gauge structure becomes non-Abelian, i.e. it leads to state changes that can be described in terms of non-commuting unitary operators acting on the input state. In the special case of non-degenerate eigensubspaces, Berry [7] showed that the state picks up a phase factor that can be described in terms of an Abelian synthetic gauge field. Close to points in parameter space where two or more energies intersect, this Abelian gauge field resembles that of a magnetic monopole; an elusive missing entity of electromagnetism. The unitary operators and phase factors are solely dependent on the adiabatic paths taken in parameter space.

2.3.1   The Wilczek–Zee Holonomy

In the adiabatic regime, the evolution of a degenerate energy eigensubspace becomes purely geometric in the sense that the state change becomes a pure function of the path in parameter space. Here, we discuss how this change can be understood as a synthetic gauge theory, as outlined by Wilczek and Zee [6].

Consider a quantum-mechanical system starting in an energetically degenerate subspace ($g_n \geq 2$) and being adiabatically driven along a path $\mathscr{C} : [0, 1] \ni s \mapsto \mathbf{R}(s)$ in an L_R-dimensional parameter space with $\mathbf{R} = (R_1, \ldots, R_{L_R})$. We shall assume $\mathscr{C}$ is a loop, i.e. $\mathbf{R}(1) = \mathbf{R}(0)$. We show that the state change in the adiabatic regime as described by (2.7) by neglecting non-adiabatic transition terms, i.e.

$$c_{n;k}(s) = c_{n;k}(0) - \sum_{l=1}^{g_n} \int_0^s ds' \langle \psi_{n,k}(\mathbf{R}(s')) | \dot{\psi}_{n,l}(\mathbf{R}(s')) \rangle c_{n;l}(s'), \quad (2.26)$$

only depends on the loop $\mathscr{C}$.

We first solve (2.26) iteratively,

$$c_{n;k}(s) = \sum_{l=1}^{g_n} \left(\delta_{kl} - \int_{\mathbf{R}(0);\mathscr{C}}^{\mathbf{R}(s)} d\mathbf{R}(s') \cdot \langle \psi_{n,k}(\mathbf{R}(s')) | \nabla_{\mathbf{R}(s)} \psi_{n,l}(\mathbf{R}(s')) \rangle \right.$$
$$+ \sum_{l'=1}^{g_n} \int_{\mathbf{R}(0);\mathscr{C}}^{\mathbf{R}(s)} d\mathbf{R}(s') \cdot \langle \psi_{n,k}(\mathbf{R}(s')) | \nabla_{\mathbf{R}(s')} \psi_{n,l'}(\mathbf{R}(s')) \rangle$$
$$\left. \times \int_{\mathbf{R}(0);\mathscr{C}}^{\mathbf{R}(s')} d\mathbf{R}(s'') \cdot \langle \psi_{n,l'}(\mathbf{R}(s'')) | \nabla_{\mathbf{R}(s'')} \psi_{n,l}(\mathbf{R}(s'')) \rangle - \cdots \right) c_{n;l}(0),$$

$$(2.27)$$

where we have used the chain rule $d/ds = \dot{\mathbf{R}}(s) \cdot \nabla_{\mathbf{R}}$. The notation $\int_{\mathbf{R}(0);\mathscr{C}}^{\mathbf{R}(s')} d\mathbf{R}(s'')\cdot$ stands for a path integral along $\mathscr{C}$, starting at $\mathbf{R}(0)$ and ending at a point $\mathbf{R}(s')$ between the end-points of $\mathscr{C}$. Since all path integrals take place over portions of $\mathscr{C}$, it follows that one can formally rewrite (2.27) at $s = 1$ as

$$c_{n;k}(1) = \sum_{l=1}^{g_n} \left(\mathscr{P} e^{i \oint_{\mathscr{C}} A_n^\alpha(\mathbf{R}) dR_\alpha} \right)_{kl} c_{n;l}(0), \quad (2.28)$$

where $\mathscr{P}$ is path ordering along $\mathscr{C}$ and sum over repeated parameter indices is understood. Here,

$$A_{n;kl}^\alpha(\mathbf{R}) = i \langle \psi_k(\mathbf{R}) | \partial^\alpha \psi_l(\mathbf{R}) \rangle \quad (2.29)$$

is the matrix-valued Hermitian synthetic gauge (Wilczek–Zee) connection with $\partial^\alpha = \partial/\partial R_\alpha$. Thus, a given input state

$$|\Phi(0)\rangle = \sum_{k=1}^{g_n} |\psi_{n;k}(\mathbf{R}(0))\rangle c_{n;k}(0) \quad (2.30)$$

transforms into

$$|\Phi(1)\rangle = e^{-iT\int_0^1 \varepsilon_n(\mathbf{R}(s))ds} \sum_{k,l=1}^{g_n} |\psi_{n;k}(\mathbf{R}(0))\rangle \left(\mathscr{P}e^{i\oint_{\mathscr{C}} A_n^\alpha(\mathbf{R})dR_\alpha}\right)_{kl} c_{n;l}(0) \quad (2.31)$$

after completing the loop. The unitary matrix $U_{n;kl}(\mathscr{C}) \equiv \left(\mathscr{P}e^{i\oint_{\mathscr{C}} A_n^\alpha(\mathbf{R})dR_\alpha}\right)_{kl}$ is the Wilczek–Zee holonomy or non-Abelian geometric phase associated with the adiabatic loop $\mathscr{C}$ in parameter space. Note that the dynamical phase factor $e^{-iT\int_0^1 \varepsilon_n(\mathbf{R}(s))ds}$ is factored out as a global phase with no observable effect on the state change.

The above synthetic gauge connection (2.29) is a quantity that we will meet throughout this book. It constitutes the basis for a synthetic gauge theory associated with adiabatic evolution. To see this, we first state the meaning of gauge choice and gauge transformations. The chosen basis of the eigensubspace is the gauge choice and a gauge transformation is a smooth local basis change,

$$|\psi_{n;k}(\mathbf{R}(s))\rangle \mapsto |\widetilde{\psi}_{n;k}(\mathbf{R}(s))\rangle = \sum_{l=1}^{g_n} |\psi_{n;l}(\mathbf{R}(s))\rangle V_{lk}(\mathbf{R}(s)), \qquad (2.32)$$

under which the synthetic gauge connection transforms as a proper gauge potential,

$$A_{n;kl}^\alpha(\mathbf{R}) \mapsto \widetilde{A}_{n;kl}^\alpha(\mathbf{R}) = i\langle\widetilde{\psi}_{n;k}(\mathbf{R})|\partial^\alpha \widetilde{\psi}_{n;l}(\mathbf{R})\rangle$$

$$= \sum_{p,q=1}^{g_n} V_{kp}^\dagger(\mathbf{R}) A_{n;pq}^\alpha(\mathbf{R}) V_{ql}(\mathbf{R}) + i\sum_{p=1}^{g_n} V_{kp}^\dagger(\mathbf{R})\partial^\alpha V_{pl}(\mathbf{R}). \quad (2.33)$$

The inhomogeneous term on the right-hand side shows that the gauge connection is not directly accessible experimentally, just as the vector potential in electro-magnetism is a gauge dependent quantity and therefore is not uniquely defined by the observable electromagnetic field. On the other hand, experimentally accessible quantities should transform unitarily under gauge transformations in order for expectation values and probabilities to be basis independent. Indeed, one finds that the Wilczek–Zee holonomy transforms unitarily,

$$U_{n;kl}(\mathscr{C}) \mapsto \sum_{p,q=1}^{g_n} V_{kp}^\dagger(\mathbf{R}(0)) U_{n;pq}(\mathscr{C}) V_{ql}(\mathbf{R}(0)) \qquad (2.34)$$

under the smooth basis change in (2.32). Similarly, the anti-symmetric curvature tensor, defined as

$$\mathsf{F}_n^{\alpha\beta}(\mathbf{R}) = \partial^\alpha A_n^\beta(\mathbf{R}) - \partial^\beta A_n^\alpha(\mathbf{R}) + i[A_n^\alpha(\mathbf{R}), A_n^\beta(\mathbf{R})], \qquad (2.35)$$

transforms unitarily under a gauge transformation. $\mathsf{F}_n^{\alpha\beta}$ is the non-Abelian analogue of the electromagnetic curvature tensor, which defines the electromagnetic field. In particular, note how the last commutator term vanishes identically in the non-degenerate case $g_n = 1$.

The curvature tensor tells us whether the gauge connection is curved or not, i.e. whether the gauge structure can have a physical influence on the adiabatic evolution or not. In particular, the curvature tensor has an immediate relation to the energy gaps of the system, as can be seen by rewriting (2.35) by means of the instantaneous Schrödinger equation $H(\mathbf{R})|\psi_{n;l}(\mathbf{R})\rangle = \varepsilon_n(\mathbf{R})|\psi_{n;l}(\mathbf{R})\rangle$ yielding

$$\mathsf{F}_{n;kl}^{\alpha\beta}(\mathbf{R})$$

$$= i \sum_{m\neq n=1}^{K} \sum_{p=1}^{g_m} \left(\frac{\langle\psi_{n;k}(\mathbf{R})|\partial^\alpha H(\mathbf{R})|\psi_{m;p}(\mathbf{R})\rangle\langle\psi_{m;p}(\mathbf{R})|\partial^\beta H(\mathbf{R})|\psi_{n;l}(\mathbf{R})\rangle}{\Delta_{nm}^2(\mathbf{R})} \right.$$

$$\left. - \frac{\langle\psi_{n;k}(\mathbf{R})|\partial^\beta H(\mathbf{R})|\psi_{m;p}(\mathbf{R})\rangle\langle\psi_{m;p}(\mathbf{R})|\partial^\alpha H(\mathbf{R})|\psi_{n;l}(\mathbf{R})\rangle}{\Delta_{nm}^2(\mathbf{R})} \right). \qquad (2.36)$$

In other words, the curvature tensor is singular where the energy gaps close. We shall see in the examples below that the vicinity regions of 'intersection points' where two or more energies cross (become degenerate) are associated with a gauge structure that resembles that of a magnetic monopole. Therefore, energy intersection points play a central role as a source of the gauge field in the space of slow parameters in the system.

Geometrically, the synthetic gauge connection can be understood in terms of a natural concept of *parallel transport* over the eigensubspaces of the slowly changing Hamiltonian. To see this, let us consider two nearby g_n-dimensional eigensubspaces $\mathscr{L}(\mathbf{R}(s))$ and $\mathscr{L}(\mathbf{R}(s+\delta s))$ along $\mathscr{C}$ with eigenprojectors $\hat{P}_n(\mathbf{R}(s))$ and $\hat{P}_n(\mathbf{R}(s+\delta s))$, respectively. Let $\mathscr{W}_n(\mathbf{R}(s)) = \{|\psi_{n;k}(\mathbf{R}(s))\rangle\}_{k=1}^{g_n}$ and $\mathscr{W}_n(\mathbf{R}(s+\delta s)) = \{|\psi_{n;k}(\mathbf{R}(s+\delta s))\rangle\}_{k=1}^{g_n}$ be two frames (ordered bases) spanning the two subspaces. These frames can be made parallel by minimising the function [8]

$$\mathscr{D}^2\left(\mathscr{W}_n(s), \mathscr{W}_n(s+\delta s)\right) = \sum_{k=1}^{g_n} \| \, |\psi_{n;k}(\mathbf{R}(s))\rangle - |\psi_{n;k}(\mathbf{R}(s+\delta s))\rangle \, \|^2$$

$$= 2g_n - 2\mathrm{ReTr}\mathscr{O}\left(\mathscr{W}_n(\mathbf{R}(s)), \mathscr{W}_n(\mathbf{R}(s+\delta s))\right) \qquad (2.37)$$

over all possible choices of frame pairs. The minimum gives a gauge invariant notion of distance between the eigensubspaces $\mathscr{L}(\mathbf{R}(s))$ and $\mathscr{L}(\mathbf{R}(s+\delta s))$, containing the Fubini–Study distance [9] as the special case for which $g_n = 1$. The key quantity in the minimisation procedure is apparently the overlap matrix

$$\mathscr{O}_{kl}\left(\mathscr{W}_n(\mathbf{R}(s)), \mathscr{W}_n(\mathbf{R}(s+\delta s))\right) = \langle\psi_{n;k}(\mathbf{R}(s))|\psi_{n;l}(\mathbf{R}(s+\delta s))\rangle \qquad (2.38)$$

between the two frames. This minimum is found by making the gauge transformation

$$|\psi_{n;l}(\mathbf{R}(s))\rangle \mapsto |\widetilde{\psi}_{n;l}(\mathbf{R}(s))\rangle = \sum_{j=1}^{g_n} |\psi_{n;j}(\mathbf{R}(s))\rangle V_{jl}(\mathbf{R}(s)) \qquad (2.39)$$

such that the corresponding overlap matrix $\langle \widetilde{\psi}_{n;k}(\mathbf{R}(s))|\widetilde{\psi}_{n;l}(\mathbf{R}(s + \delta s))\rangle$ becomes positive definite to first order in δs. This yields the *connection* (rule for parallel transport)

$$\dot{V}(\mathbf{R}(s))V^{\dagger}(\mathbf{R}(s)) = i A_n^{\alpha}(\mathbf{R}(s))\dot{R}_{\alpha}. \qquad (2.40)$$

Thus, the synthetic gauge connection $A_n^{\alpha}(\mathbf{R}(s))$ determines the rule for parallel transport over the eigensubspaces along the path $\mathscr{C}$ in parameter space. The Wilczek–Zee unitary $U_n(\mathscr{C})$ is thereby the holonomy arising when the system is parallel transported around a loop.

2.3.2 Adiabatic Evolution of a Tripod

To realise the non-Abelian property of the Wilczek–Zee holonomy, we need a system with a symmetry-protected degeneracy over the relevant part of parameter space. As we will discuss in more detail in Sect. 5.4, this can be found in the context of laser-controlled atomic systems exhibiting a tripod structure that consists of three 'ground state' energy levels $|g_1\rangle, |g_2\rangle, |g_3\rangle$ coupled by three laser fields to one and the same excited state $|e\rangle$, see Fig. 2.2. We illustrate the Wilczek–Zee holonomy along loops in parameter space of a tripod.

By employing the rotating wave approximation (see Sect. 6.1) in the interaction picture, we obtain the tripod Hamiltonian

$$\hat{H}_{\text{tripod}} = \delta_1|g_1\rangle\langle g_1| + \delta_2|g_2\rangle\langle g_2| + \delta_3|g_3\rangle\langle g_3|$$
$$+ \Omega \left(\omega_1|e\rangle\langle g_1| + \omega_2|e\rangle\langle g_2| + \omega_3|e\rangle\langle g_3| + \text{H.c.} \right) \qquad (2.41)$$

with detunings $\delta_j = 2\pi\nu_j - f_{je}$, ν_j and f_{je} being the field frequencies and energy spacings, respectively, and $\omega_j = \omega_j(s)$ being complex-valued parameters describing the phase and amplitude of the fields. We further assume that $\sum_j |\omega_j|^2 = 1$, which means that Ω measures the overall strength of the system–laser interaction.

We first look for restrictions on the parameters that generate a degenerate pair of energy eigenstates of the form $c_1|g_1\rangle + c_2|g_2\rangle + c_3|g_3\rangle$. These are called dark states as they do not involve the potentially unstable excited state $|e\rangle$. Given this form, the

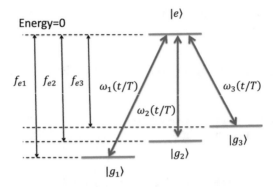

Fig. 2.2 Tripod system consisting of three 'ground state' levels $|g_j\rangle$, $j = 1, 2, 3$, coupled to an excited state $|e\rangle$ by three laser fields. $-f_{ej}$ are the bare energies of the ground state levels relative the energy of the excited state. The slowly varying complex-valued parameters $\omega_j(t/T)$ describe the amplitude and phase of the lasers and define the parameter space of the system. If the laser frequencies are chosen so as to give rise to the same detuning, the system exhibits a doubly degenerate dark eigensubspace that traces out a path in Span$\{|g_1\rangle, |g_2\rangle, |g_3\rangle\}$ under adiabatic change of $\omega_j(t/T)$

eigenvalue equation for $\hat{H}_{\text{tripod}}$ gives

$$\omega_1 c_1 + \omega_2 c_2 + \omega_3 c_3 = 0,$$

$$\delta_j c_j = \varepsilon_d c_j, \quad j = 1, 2, 3, \qquad (2.42)$$

with ε_d the energy of the dark state subspace. These equations have precisely two solution if and only if $\delta_j = \varepsilon_d$. Thus, there is a degenerate pair of dark energy eigenstates $|D_1(\mathbf{R})\rangle$ and $|D_2(\mathbf{R})\rangle$, for all $\mathbf{R} = (\omega_1, \omega_2, \omega_3)$, with energy ε_d being the common detuning of the three laser fields. Explicitly, by defining $\omega_1 = e^{-iS_1} \sin\theta \cos\phi$, $\omega_2 = e^{-iS_2} \sin\theta \sin\phi$, and $\omega_3 = \cos\theta$, we may take

$$|D_1(\mathbf{R})\rangle = -e^{iS_1} \sin\phi|g_1\rangle + e^{iS_2} \cos\phi|g_2\rangle,$$

$$|D_2(\mathbf{R})\rangle = e^{iS_1} \cos\theta \cos\phi|g_1\rangle + e^{iS_2} \cos\theta \sin\phi|g_2\rangle - \sin\theta|g_3\rangle, \quad (2.43)$$

where $\theta \in [0, \pi]$. In addition, there are two non-degenerate bright states $|B_\pm(\mathbf{R})\rangle$ with energies $\frac{1}{2}\left(\varepsilon_d \pm \sqrt{\varepsilon_d^2 + 4\Omega^2}\right)$. With $\hat{P}_d(\mathbf{R}) = |D_1(\mathbf{R})\rangle\langle D_1(\mathbf{R})| + |D_2(\mathbf{R})\rangle\langle D_2(\mathbf{R})|$, we can thus write

$$\hat{H}_{\text{tripod}} = \varepsilon_d \hat{P}_d(\mathbf{R}) + \frac{1}{2}\left(\varepsilon_d + \sqrt{\varepsilon_d^2 + 4\Omega^2}\right)|B_+(\mathbf{R})\rangle\langle B_+(\mathbf{R})|$$

$$+ \frac{1}{2}\left(\varepsilon_d - \sqrt{\varepsilon_d^2 + 4\Omega^2}\right)|B_-(\mathbf{R})\rangle\langle B_-(\mathbf{R})|. \qquad (2.44)$$

Now we assume that $\mathscr{C} : [0, 1] \ni s \mapsto \mathbf{R} = \mathbf{R}(s)$ is a smooth loop in the four-dimensional parameter space $\mathbf{R} = (\theta, \phi, S_1, S_2)$ and that $\sqrt{\varepsilon_d^2 + \Omega^2} \neq 0$. In the adiabatic regime, the run-time T is so large that transitions between the dark subspace and the two bright states become negligible. When this condition is satisfied, the loop $\mathscr{C}$ in the space of slowly changing $\mathbf{R}$ approximately determines the non-trivial action of the time-evolution operator. To illustrate this we consider two different loops $\mathscr{C}_{(i)}$ and $\mathscr{C}_{(ii)}$, leading to two non-commuting Wilczek–Zee holonomies $\hat{U}(\mathscr{C}_{(i)})$ and $\hat{U}(\mathscr{C}_{(ii)})$. The first loop is

$$\mathscr{C}_{(i)} : [0, 1] \ni s \mapsto \mathbf{R}_{(i)}(s) = (\theta(s), \phi(s), 0, 0),$$

$$\theta(1) = \theta(0) = 0, \quad \phi(1) = \phi(0) = 0, \text{ mod } (2\pi), \qquad (2.45)$$

yielding the operator-valued gauge connection

$$\hat{A}^\phi_{(i)}(\theta, \phi) = \hat{\sigma}_y \cos\theta, \quad \hat{A}^\theta_{(i)}(\theta, \phi) = 0 \qquad (2.46)$$

with

$$\hat{\sigma}_y = -i|g_2\rangle\langle g_1| + i|g_1\rangle\langle g_2|. \qquad (2.47)$$

We thus find

$$\hat{U}(\mathscr{C}_{(i)}) = e^{-i\hat{\sigma}_y \oint_{\mathscr{C}_{(i)}} \cos\theta d\phi}. \qquad (2.48)$$

The second loop is

$$\mathscr{C}_{(ii)} : [0, 1] \ni s \mapsto \mathbf{R}_{(ii)}(s) = \left(\theta(s), \frac{\pi}{2}, 0, S_2(s)\right),$$

$$\theta(1) = \theta(0) = 0, \quad S_2(1) = S_2(0) = 0, \text{ mod } (2\pi), \qquad (2.49)$$

yielding instead the operator-valued gauge connection

$$\hat{A}^{S_2}_{(ii)}(\theta, \phi) = -\hat{P}_2 \cos\theta, \quad \hat{A}^\theta_{(ii)}(\theta, \phi) = 0 \qquad (2.50)$$

with

$$\hat{P}_2 = |g_2\rangle\langle g_2|. \qquad (2.51)$$

We thus find

$$\hat{U}(\mathscr{C}_{(ii)}) = e^{i\hat{P}_2 \oint_{\mathscr{C}_{(ii)}} \cos\theta d S_2}. \qquad (2.52)$$

We note that the two unitary operators $\hat{U}(\mathscr{C}_{(i)})$ and $\hat{U}(\mathscr{C}_{(ii)})$ do not commute, which demonstrates the non-Abelian property of the Wilczek–Zee holonomy. The target space of both $\hat{U}(\mathscr{C}_{(i)})$ and $\hat{U}(\mathscr{C}_{(ii)})$ is Span$\{|g_1\rangle, |g_2\rangle\}$, which defines the state space of a single qubit. Due to the non-Abelian feature of these Wilczek–Zee holonomies, they can be used to manipulate the state of this qubit in an arbitrary fashion. This is the idea of holonomic quantum computation [10, 11].

The integrals $\oint_{\mathscr{C}_{(i)}} \cos\theta d\phi$ and $\oint_{\mathscr{C}_{(ii)}} \cos\theta dS_2$ are our first examples of monopole structures. These monopoles both sit at the energy intersection point $\varepsilon_d = \Omega = 0$ (cf. (2.36)), in the Cartesian subspaces $\mathbf{R}_{(i)} = \sqrt{\varepsilon_d^2 + \Omega^2}(\sin\theta\cos\phi, \sin\theta\sin\phi, \cos\theta)$ and $\mathbf{R}_{(ii)} = \sqrt{\varepsilon_d^2 + \Omega^2}(\sin\theta\cos S_2, \sin\theta\sin S_2, \cos\theta)$. Monopoles associated with intersection points are a recurrent theme of these Lecture Notes. In the next section, we shall therefore discuss in more detail the 'canonical' monopole, which is of magnetic-type and arises in the Abelian context.

2.3.3 Closing the Energy Gap: Abelian Magnetic Monopole in Adiabatic Evolution

The synthetic gauge structure associated with adiabatic evolution, as outlined above, is similar to the gauge structure in the theories of elementary particles. For instance, in the case of adiabatic systems with non-degenerate energy eigenstates, this structure is an exact analog of that of an electromagnetic field. Nevertheless, the synthetic gauge theory is not constrained by the empirical content of these other gauge theories, and new phenomena can therefore occur. Here, we demonstrate the appearance of magnetic monopoles in adiabatic evolution of non-degenerate states, which has been an elusive entity in electromagnetism.

The non-degenerate case was examined by Berry in his seminal work on geometric phase factors in adiabatic evolution [7]. Here, the synthetic gauge connection is no longer matrix-valued and the holonomy reduces to a phase factor,

$$e^{i\gamma_n(\mathscr{C})} = e^{i\oint_{\mathscr{C}} A_n^\alpha(\mathbf{R})dR_\alpha} \tag{2.53}$$

which defines the Berry phase

$$\gamma_n(\mathscr{C}) = \oint_{\mathscr{C}} A_n^\alpha(\mathbf{R})dR_\alpha, \mod(2\pi), \tag{2.54}$$

of the loop. Here,

$$A_n^\alpha(\mathbf{R}) = i\langle\psi_n(\mathbf{R})|\partial^\alpha\psi_n(\mathbf{R})\rangle \tag{2.55}$$

is an Abelian synthetic gauge connection, sometimes called a *Berry connection*. Note that path ordering is not needed since the holonomy is Abelian ($[A_n(\mathbf{R}), A_n(\mathbf{R}')] = 0$ for any pair $\mathbf{R}$ and $\mathbf{R}'$ along the path) in the $g_n = 1$ case. If the system starts in the eigenstate $|\psi_n(\mathbf{R}(0))\rangle$ at $s = 0$, it ends up approximately in the state

$$|\Phi(1)\rangle = e^{-i \int_0^1 \varepsilon_n(\mathbf{R}(s))ds + i\gamma_n(\mathscr{C})}|\Phi(0)\rangle, \tag{2.56}$$

which shows that the cyclic phase $\kappa_n(1)$ defined in (2.4) is the sum of a dynamical and a geometric part in the adiabatic limit $T \to \infty$.

Furthermore, the gauge transformation reduces to a phase transformation

$$|\psi_n(\mathbf{R})\rangle \mapsto |\widetilde{\psi}(\mathbf{R})\rangle = e^{if(\mathbf{R})}|\psi_n(\mathbf{R})\rangle, \tag{2.57}$$

being local in parameter space. Under such a phase change, the gauge connection transforms as

$$A_n^\alpha(\mathbf{R}) \mapsto \widetilde{A}_n^\alpha(\mathbf{R}) = A_n^\alpha(\mathbf{R}) - \partial^\alpha f(\mathbf{R}), \tag{2.58}$$

while the Berry phase factor $e^{i\gamma_n(\mathscr{C})}$ and the Abelian curvature tensor $\mathsf{F}_n^{\alpha\beta}(\mathbf{R}) = \partial^\alpha A_n^\beta(\mathbf{R}) - \partial^\beta A_n^\alpha(\mathbf{R})$ both are unchanged.

We now demonstrate the significance of the Berry phase in the case of a spin$-S$ and show how a magnetic monopole appears in adiabatic evolution of this system. Let $\hat{\mathbf{S}} = (\hat{S}_x, \hat{S}_y, \hat{S}_z)$ be the spin operator and $\mathbf{B} = B(\sin\theta\cos\phi, \sin\theta\sin\phi, \cos\theta) \equiv B\mathbf{n}$ a slowly changing magnetic field. We use the standard notation $S_z|S, M\rangle = M|S, M\rangle$, $M = -S, \ldots, S$ ($\hbar = 1$ as before) with $S(S + 1)$ being the eigenvalue of $\mathbf{S}^2$. Suppose the spin carries a magnetic moment μ and interacts with the magnetic field according to the Zeeman Hamiltonian

$$\hat{H}(\mathbf{B}) = -\mu\hat{\mathbf{S}} \cdot \mathbf{B} = -\mu\hat{U}(\theta, \phi)\hat{S}_z\hat{U}^\dagger(\theta, \phi), \tag{2.59}$$

where

$$\hat{U}(\theta, \phi) = e^{-i\phi\hat{S}_z}e^{-i\theta\hat{S}_y}e^{i\phi\hat{S}_z} \tag{2.60}$$

is the rotation operator. The instantaneous energy eigenstates are spin eigenstates $|S, M; \mathbf{n}\cdot\mathbf{S}\rangle = \hat{U}(\theta, \phi)|S, M\rangle$ and with corresponding energy eigenvalues $-M\mu B$. Clearly, the magnetic field strength determines the size of the energy gaps.

With the above choice of rotation operators, we have $\hat{U}(0, \phi) = \hat{1}$ and $\hat{U}(\pi, \phi) = e^{-i\pi\hat{S}_y}e^{2i\phi\hat{S}_z}$. This entails that the corresponding energy eigenvectors cannot be unique simultaneously at the north and south pole. For example, by choosing the phase of the eigenvectors $|\psi_{M;z}\rangle$ of $\hat{H}_z \equiv \hat{H}(B\mathbf{e}_z)$ to be independent of ϕ, for instance, by putting $|\psi_{M;z}\rangle = |S, M\rangle$, the resulting instantaneous eigenvectors $\hat{U}(\theta, \phi)|\psi_{M;z}\rangle$ are unique at the north pole but yields a singular gauge potential at the south pole. For the same reference eigenvectors, one may move this singularity to the north pole by instead choosing the rotation operators $\widetilde{\widehat{U}}(\theta, \phi) = e^{-i\phi \hat{S}_z} e^{-i\theta \hat{S}_y} e^{-i\phi \hat{S}_z}$. On the other hand, the *gauge section* [12]

$$\mathscr{S} = \{\hat{U}(\theta, \phi), \theta \in [0, \pi); \widetilde{\widehat{U}}(\theta, \phi), \theta \in (0, \pi]\} \tag{2.61}$$

is globally well-defined for any choice of eigenvectors of $\hat{H}_z$. This single-valued section captures the monopole structure corresponding to the $2S+1$ fold degeneracy at $\mathbf{B} = 0$, the *conical intersection* (CI) point of the system.

Let us now calculate the gauge connection by using $\hat{U}(\theta, \phi)$ and $\widetilde{\widehat{U}}(\theta, \phi)$. We find the non-zero components,

$$A_M^\phi(\theta, \phi) = -M(1 - \cos\theta),$$
$$\widetilde{A}_M^\phi(\theta, \phi) = M(1 + \cos\theta) \tag{2.62}$$

with $dR_\phi = d\phi$. We obtain the Berry phase expressions

$$\gamma_M(\mathscr{C}) = -M \oint_{\mathscr{C}} (1 - \cos\theta) d\phi = -M\Omega(\mathscr{C}),$$
$$\widetilde{\gamma}_M(\mathscr{C}) = M \oint_{\mathscr{C}} (1 + \cos\theta) d\phi = -M\Omega(\mathscr{C}) + 2M \oint_{\mathscr{C}} d\phi, \tag{2.63}$$

$\Omega(\mathscr{C})$ being the solid angle enclosed by the loop $\mathscr{C}$. These Berry phases thus differ by an integer multiple of 2π (since $\oint_{\mathscr{C}} d\phi$ is the winding number around the z-axis of parameter space) and therefore correspond to the same Berry phase factor $e^{i\widetilde{\gamma}_M(\mathscr{C})} = e^{i\gamma_M(\mathscr{C})}$. Equivalently, we may note that the above gauge connections differ by a pure single-valued gauge

$$\widetilde{A}_M^\phi(\theta, \phi) - A_M^\phi(\theta, \phi) = 2M \tag{2.64}$$

in the overlapping region $\theta \in (0, \pi)$, thus giving rise to a unique curvature tensor

$$\mathsf{F}_M^{\theta, \phi} = \partial^\theta A_M^\phi - \partial^\phi A_M^\theta = -M\sin\theta, \tag{2.65}$$

which describes a gauge field of a magnetic monopole sitting at the CI at $\mathbf{B} = 0$ of parameter space. The monopole becomes explicit by calculating the synthetic 'magnetic field' $\mathbf{B}^{(M)}$, defined as the curl of the synthetic vector potential,

$$\mathbf{A}^{(M)}(\mathbf{B}) = i\langle S, M|U^\dagger(\theta, \phi) \left(\mathbf{e}_B \frac{\partial}{\partial B} + \mathbf{e}_\theta \frac{1}{B} \frac{\partial}{\partial \theta} + \mathbf{e}_\phi \frac{1}{B\sin\theta} \frac{\partial}{\partial \phi}\right) U(\theta, \phi)|S, M\rangle$$
$$= -M \frac{(1 - \cos\theta)}{B\sin\theta} \mathbf{e}_\phi. \tag{2.66}$$

This gives the field

$$\mathbf{B}^{(M)}(\mathbf{B}) = \nabla_{\mathbf{B}} \times \mathbf{A}^{(M)}(\mathbf{B}) = \frac{\mathbf{B}}{B^2} \frac{1}{\sin\theta} \frac{\partial}{\partial\theta} \left(\sin\theta \mathbf{A}_{M;\phi}\right) + \cdots = -M\frac{\mathbf{B}}{B^3} \quad (2.67)$$

corresponding to a monopole of strength $-M$ sitting at the CI at $\mathbf{B} = \mathbf{0}$. Note that any other gauge choice, such as the one obtained from $\tilde{\widetilde{U}}(\theta, \phi)$, would give the same synthetic magnetic field. Stokes' theorem implies that the Berry phase is the flux of the monopolar field in (2.67) through any surface enclosed by $\mathscr{C}$.

2.4 Born–Oppenheimer Theory

As we have seen, the notion of adiabaticity plays a vital role in time-independent systems. It can be used as a simplifying factor when solving the time-independent Schrödinger equation of composite quantum-mechanical systems consisting of 'fast' and 'slow' degrees of freedom, such as the electrons and nuclei of a molecule. Here, 'fast' and 'slow' are related to a large difference in mass between the two types of subsystems. This is the idea behind the *Born–Oppenheimer approximation*, in which the wave function is approximated as a product of a 'fast' and a 'slow' part. This approximation is based, in turn, on restricting the state space of the fast system to a subspace that is parametrically dependent on the coordinates of the slow system. The parameter dependence of this subspace is the origin of a synthetic gauge structure arising in the Born–Oppenheimer theory for such composite systems [13].

2.4.1 Synthetic Gauge Structure of Born–Oppenheimer Theory

The starting point of Born–Oppenheimer theory is a system with $L_q + L_Q$ quantum-mechanical degrees of freedom, which we divide into $\hat{\mathbf{q}} = (\hat{q}^1, \ldots, \hat{q}^{L_q})$ and $\hat{\mathbf{Q}} = (\hat{Q}^1, \ldots, \hat{Q}^{L_Q})$, thereby defining the configuration space of the two subsystems. We wish to solve the time-independent Schrödinger equation

$$\hat{H}|\Psi\rangle = E|\Psi\rangle \quad (2.68)$$

with the system Hamiltonian

$$\hat{H} = -\frac{1}{2m_\alpha} \hat{P}_\alpha \hat{P}^\alpha - \frac{1}{2\mu_a} \hat{p}_a \hat{p}^a + V(\hat{\mathbf{q}}, \hat{\mathbf{Q}}), \quad (2.69)$$

where $\hat{P}^\alpha$ and $\hat{p}^a$ are conjugate momenta of $\hat{Q}^\alpha$ and $\hat{q}^a$, respectively, and we sum over repeated tensor indices α, a. The first two terms on the right-hand side of this expression constitute the kinetic energy of the system and V is the potential energy describing the interactions within the system as well as interactions with any external systems. m_α and μ_a are the masses associated with Q_α and q_a, respectively. The key assumption is that the smallest m_α is much larger than the largest μ_a. In this sense, $\hat{\mathbf{Q}}$ and $\hat{\mathbf{q}}$ represent the slow and fast subsystem, respectively, based on the fact that systems with large mass move slower than systems with small mass.

To solve (2.68) is generally a formidable task, but one may use the difference in masses in order to make simplifying approximations. Here, we outline the basis for these approximations.

We begin by defining

$$\hat{h}(\mathbf{Q}) = -\frac{1}{2\mu_a}\hat{p}_a\hat{p}^a + V(\hat{\mathbf{q}}, \mathbf{Q}),\tag{2.70}$$

being the Hamiltonian of the fast subsystem for which $\mathbf{Q}$ plays the role of a set of parameters. We assume that we can restrict, for each $\mathbf{Q}$, the Hilbert space $\mathscr{H}_q$ of the fast subsystem to an N-dimensional subspace $\mathscr{L}_q(\mathbf{Q})$, which in practise is a proper subspace of $\mathscr{H}_q$. For simplicity, we assume N does not change over $\mathbf{Q}$ and let $\{|l(\mathbf{Q})\rangle\}_{l=1}^N$ span $\mathscr{L}_q(\mathbf{Q})$. Given this truncated $\mathbf{Q}$-dependent basis, we use the ansatz

$$|\widetilde{\Psi}(\mathbf{Q})\rangle = \sum_{l=1}^N |l(\mathbf{Q})\rangle \chi_l(\mathbf{Q}) \equiv \langle \mathbf{Q}|\widetilde{\Psi}\rangle \tag{2.71}$$

to find an approximate solutions $|\widetilde{\Psi}\rangle \approx |\Psi\rangle$ and $\widetilde{E} \approx E$ by solving the equation

$$\int d\mathbf{Q}' \langle \mathbf{Q}|\hat{H}|\mathbf{Q}'\rangle |\widetilde{\Psi}(\mathbf{Q}')\rangle = \widetilde{E}|\widetilde{\Psi}(\mathbf{Q})\rangle,\tag{2.72}$$

which is (2.68) in position representation, restricted to the subspace $\mathscr{L}_q(\mathbf{Q})$ of $\mathscr{H}_q$. Since $\langle \mathbf{Q}|\hat{P}_\alpha\hat{P}^\alpha|\mathbf{Q}'\rangle = -\delta(\mathbf{Q} - \mathbf{Q}')\partial_\alpha\partial^\alpha$ with $\partial^\alpha = \partial/\partial Q^\alpha$, one can perform the integration over $\mathbf{Q}'$ yielding

$$\left(-\frac{1}{2m_\alpha}\partial_\alpha\partial^\alpha + \hat{h}(\mathbf{Q})\right)|\widetilde{\Psi}(\mathbf{Q})\rangle = \widetilde{E}|\widetilde{\Psi}(\mathbf{Q})\rangle.\tag{2.73}$$

By multiplying with $\langle k(\mathbf{Q})|$, one obtains the coupled equations

$$-\frac{1}{2m_\alpha}\sum_{l,j=1}^N \left(\delta_{kj}\partial_\alpha - i A_{\alpha;kj}(\mathbf{Q})\right)\left(\delta_{jl}\partial^\alpha - i A_{jl}^\alpha(\mathbf{Q})\right)\chi_l(\mathbf{Q})$$

$$+\sum_{l=1}^N \left(\mathscr{V}_{kl}(\mathbf{Q}) + \langle k(\mathbf{Q})|\hat{h}(\mathbf{Q})|l(\mathbf{Q})\rangle\right)\chi_l(\mathbf{Q}) = \widetilde{E}\chi_k(\mathbf{Q})\tag{2.74}$$

for the expansion coefficients $\chi_k(\mathbf{Q})$. Here,

$$A^\alpha_{kl}(\mathbf{Q}) = i\langle k(\mathbf{Q})|\partial^\alpha l(\mathbf{Q})\rangle \tag{2.75}$$

and

$$\mathscr{V}_{kl}(\mathbf{Q}) = \frac{1}{2m_\alpha}\langle \partial_\alpha k(\mathbf{Q})|\left(\hat{1} - \sum_{j=1}^{N}|j(\mathbf{Q})\rangle\langle j(\mathbf{Q})|\right)|\partial^\alpha l(\mathbf{Q})\rangle \tag{2.76}$$

are the synthetic gauge connection and scalar gauge potential, respectively. Note that the projector $\hat{1} - \sum_{j=1}^{N}|j(\mathbf{Q})\rangle\langle j(\mathbf{Q})|$ is non-zero if $N < \dim \mathscr{H}_q$; it projects onto the orthogonal complement of $\mathscr{L}_q(\mathbf{Q})$.

It should be emphasised that the crucial assumption we have made in order to allow the synthetic gauge quantities in (2.75) and (2.76) to be non-trivial is that $\mathscr{L}_q(\mathbf{Q})$ is a proper subspace of the Hilbert space. On the other hand, if $\mathscr{L}_q(\mathbf{Q})$ had spanned $\mathscr{H}_q$, then one could use a $\mathbf{Q}$-independent basis $\{|k\rangle\}$, which trivially gives rise to vanishing A and $\mathscr{V}$. However, in the practically relevant case where $\mathscr{L}_q(\mathbf{Q})$ is a proper subspace of $\mathscr{H}_q$, we can expect the synthetic gauge structure to give rise to a physical effect when solving the coupled equations in (2.74). Before demonstrating this, we discuss two important basis choices in Born–Oppenheimer theory in the next section and how the large mass difference between the fast and slow subsystems gives rise to the Born–Oppenheimer approximation (Sect. 2.4.3).

2.4.2 Adiabatic Versus Diabatic Representations

In the previous section, we demonstrated how the time-independent Schrödinger equation can be rewritten as a finite set of coupled differential equations, given a basis spanning a $\mathbf{Q}$-dependent subspace of the Hilbert space of the fast subsystem. We now consider two special choices of basis: the adiabatic and the diabatic representation. We shall see that the latter only exists if a certain condition is satisfied. This condition relates to the synthetic gauge structure of Born–Oppenheimer theory in a subtle way.

Let us first consider the adiabatic basis. This basis is the truncated part of the eigenbasis of $h(\mathbf{Q})$,

$$\hat{h}(\mathbf{Q})|\psi_n(\mathbf{Q})\rangle = \varepsilon_n(\mathbf{Q})|\psi_n(\mathbf{Q})\rangle. \tag{2.77}$$

The $\mathbf{Q}$-dependent eigenvalues $\varepsilon_n(\mathbf{Q})$ form *adiabatic potential surfaces* (APS) for the slow degrees of motion. The existence of an adiabatic basis is assured by the existence of orthonormal solutions of the eigenvalue equation of $\hat{h}(\mathbf{Q})$.

Next, we turn to the diabatic case. A diabatic basis is defined as a set of states of the fast system spanning $\mathscr{L}_q(\mathbf{Q})$ for which all the momentum coupling terms can be transformed away for all $\mathbf{Q}$. These states are generally not eigenstates of the Hamiltonian in (2.70) of the fast subsystem, meaning that the diabatic and adiabatic bases are generally different. The question is: if we have a set of states of the fast subsystem that accurately describes the physical situation, is it possible to find a unitary transformation such that the new basis is diabatic? As noted above, if the basis is complete, we can choose a $\mathbf{Q}$-independent basis for which all momentum couplings are trivially zero thus making the basis diabatic. However, in practise one is limited to a small (or at least finite) number of states of the fast subsystem, thus only a subspace of the full Hilbert space is needed. In this case it is impossible to find a diabatic basis, except in very rare cases [14]. The argument goes as follows:

Suppose we have a set of states $\{|k(\mathbf{Q})\rangle\}_{k=1}^{N}$ of the fast subsystem, defining the gauge connection $A_{kl}^{\alpha} = i\langle k(\mathbf{Q})|\partial^{\alpha}l(\mathbf{Q})\rangle$. We wish to find a local, i.e. $\mathbf{Q}$-dependent, unitary transformation

$$|k(\mathbf{Q})\rangle \longrightarrow |\widetilde{k}(\mathbf{Q})\rangle = \sum_{l=1}^{N} |l(\mathbf{Q})\rangle V_{lk}(\mathbf{Q}) \tag{2.78}$$

such that $\langle \widetilde{k}(\mathbf{Q})|\partial^{\alpha}\widetilde{l}(\mathbf{Q})\rangle = 0$. This leads to the following equations for the unitary transformation

$$i\partial^{\alpha} V_{kl}(\mathbf{Q}) = \sum_{j=1}^{N} A_{kj}^{\alpha}(\mathbf{Q}) V_{jl}(\mathbf{Q}). \tag{2.79}$$

Solutions to this exist if and only if the 'curl' of the right-hand side is zero, i.e.

$$\partial^{\alpha}\left(\sum_{j=1}^{N} A_{kj}^{\beta}(\mathbf{Q}) V_{jl}(\mathbf{Q})\right) - \partial^{\beta}\left(\sum_{j=1}^{N} A_{kj}^{\alpha}(\mathbf{Q}) V_{jl}(\mathbf{Q})\right) = 0. \tag{2.80}$$

By using (2.79), this can be rewritten in the more compact form

$$F^{\alpha\beta}(\mathbf{Q}) = 0, \tag{2.81}$$

where we recognise the curvature tensor

$$F^{\alpha\beta}(\mathbf{Q}) = \partial^{\alpha} A^{\beta}(\mathbf{Q}) - \partial^{\alpha} A^{\beta}(\mathbf{Q}) + i[A^{\alpha}(\mathbf{Q}), A^{\beta}(\mathbf{Q})] \tag{2.82}$$

of the $\mathbf{Q}$-dependent subspace $\mathscr{L}_q(\mathbf{Q})$. Thus, a diabatic representation is only possible if $\mathscr{L}_q(\mathbf{Q})$ is flat over parameter space. This is a very rare situation in practise. It basically occurs only for one-dimensional slow systems, such as in the vibrations of a diatomic molecule, for which (2.81) is trivially satisfied.

2.4.3 Born–Oppenheimer Approximation

The observant reader may have noted that the discussion above has not made use yet of the mass difference of the 'fast' and 'slow' subsystems. All results up to this point are valid, in principle, for any subdivision of the degrees of freedom of the system. Now, we shall make use of the mass difference and see how it leads to the Born–Oppenheimer approximation, which is the time-independent counterpart to the adiabatic approximation for slowly varying time-dependent systems discussed in Sects. 2.2 and 2.3.

The Born–Oppenheimer approximation is applicable whenever a state

$$|\widetilde{\Psi}(\mathbf{Q})\rangle = \chi_n(\mathbf{Q})|\psi_n(\mathbf{Q})\rangle \tag{2.83}$$

of the composite system in position representation of the slow subsystem, approximates well the exact solution $\langle \mathbf{Q}|\Psi\rangle$ of (2.68). Here, $|\psi_n(\mathbf{Q})\rangle$ is assumed to be a non-degenerate eigenstate of the Hamiltonian $h(\mathbf{Q})$ of the fast subsystem. In this case, we obtain from (2.74) the decoupled equation

$$-\frac{1}{2m_\alpha}\left(\partial_\alpha - i A_{n;\alpha}(\mathbf{Q})\right)\left(\partial^\alpha - i A_n^\alpha(\mathbf{Q})\right)\chi_n(\mathbf{Q})$$

$$+\left(\mathcal{V}_n(\mathbf{Q}) + \varepsilon_n(\mathbf{Q})\right)\chi_n(\mathbf{Q}) = \widetilde{E}\chi_n(\mathbf{Q}), \tag{2.84}$$

which is a Schrödinger-like equation describing a system in an Abelian gauge field $\partial^\alpha A_n^\beta(\mathbf{Q}) - \partial^\beta A_n^\alpha(\mathbf{Q})$ and moving on the modified APS $\mathcal{V}_n(\mathbf{Q}) + \varepsilon_n(\mathbf{Q})$. The extra scalar contribution

$$\mathcal{V}_n(\mathbf{Q}) = \frac{1}{2m_\alpha}\sum_{n'\neq n}\langle\partial_\alpha\psi_n(\mathbf{Q})|\psi_{n'}(\mathbf{Q})\rangle\langle\psi_{n'}(\mathbf{Q})|\partial^\alpha\psi_n(\mathbf{Q})\rangle \tag{2.85}$$

originates solely from the $\mathbf{Q}$ dependence of the eigenstate $|\psi_n(\mathbf{Q})\rangle$.

The Born–Oppenheimer approximation can be justified by noting that the off-diagonal ($n' \neq n$) gauge couplings

$$A_{nn'}^\alpha = i\frac{\langle\psi_n(\mathbf{Q})|\partial^\alpha\hat{h}(\mathbf{Q})|\psi_{n'}(\mathbf{Q})\rangle}{\varepsilon_{n'}(\mathbf{Q}) - \varepsilon_n(\mathbf{Q})} \tag{2.86}$$

contain the energy gaps in the denominator. These gaps are typically large relative the differences between the eigenvalues $\widetilde{E}$ obtained by solving (2.84). This can be understood by noting that the ratio of kinetic energies of the fast and slow system roughly scales as the inverse ratio of their typical masses, from which it follows that the kinetic energy of the fast subsystem is much larger than the kinetic energy of the slow subsystem. In this way, one can expect that these gauge couplings are small and can be ignored relative the diagonal terms. By the same reasoning, the off-diagonal terms of the scalar gauge potential $\mathcal{V}_{nn'}(\mathbf{Q})$ can be neglected and one thereby arrives at the Schrödinger-like equation (2.84) for each expansion coefficient.

Clearly the above argument breaks down if two or more APSs cross. These points are at the same time central for having a synthetic gauge structure that has a non-trivial effect on the slow system evolution. In other words, we are interested in the slow motion in the presence of but sufficiently far away from points where APSs cross. In next section, we discuss an illustrative physical setting where this happens.

2.4.4 Synthetic Gauge Structure of an Atom in an Inhomogeneous Magnetic Field

Suppose an atom with centre of mass at $\mathbf{x}$ and angular momentum J associated with a magnetic dipole moment μ is moving in a time-independent inhomogeneous magnetic field $\mathbf{b}(\mathbf{x})$. For simplicity, we assume $J = \frac{1}{2}$, which means that we have two Born–Oppenheimer channels corresponding to the two local angular momentum projections $M = \pm\frac{1}{2}$ along the magnetic field direction at each $\mathbf{x}$. The Hamiltonian describing the system is

$$\hat{H} = -\frac{\nabla_{\mathbf{x}}^2}{2m} + \mu\hat{\mathbf{J}} \cdot \mathbf{b}(\mathbf{x}) = -\frac{\nabla_{\mathbf{x}}^2}{2m} + \mu b(\mathbf{x})\hat{\mathbf{J}} \cdot \mathbf{n}(\mathbf{x}), \qquad (2.87)$$

where $\mathbf{n}(\mathbf{x}) = (\sin\vartheta(\mathbf{x})\cos\varphi(\mathbf{x}), \sin\vartheta(\mathbf{x})\sin\varphi(\mathbf{x}), \cos\vartheta(\mathbf{x}))$ is a unit vector describing the local direction of the magnetic field at $\mathbf{x}$. Here, the fast and slow degrees of freedom are the angular momentum and the centre of mass, respectively, of the atom. The Hamiltonian of the fast subsystem is $\hat{h}(\mathbf{x}) = \mu b(\mathbf{x})\hat{\mathbf{J}} \cdot \mathbf{n}(\mathbf{x})$, thus the slow parameter is the position $\mathbf{x} \in \mathbb{R}^3$.

To find the synthetic potentials, the unitary operator

$$\hat{U}(\vartheta(\mathbf{x}), \varphi(\mathbf{x})) = e^{-i\varphi(\mathbf{x})\hat{J}_z} e^{-i\vartheta(\mathbf{x})\hat{J}_y} e^{i\varphi(\mathbf{x})\hat{J}_z} \qquad (2.88)$$

that rotates the reference states $|\pm\rangle = |\frac{1}{2}, \pm\frac{1}{2}\rangle$ so that it points along the local direction $\mathbf{n}(\mathbf{x})$ of the magnetic field is useful. One finds the synthetic vector and scalar gauge potentials

$$\mathbf{A}^{(\pm)}(\mathbf{x}) = i\langle\pm|\hat{U}^\dagger(\vartheta(\mathbf{x}), \varphi(\mathbf{x}))\nabla_{\mathbf{x}}\hat{U}(\vartheta(\mathbf{x}), \varphi(\mathbf{x}))|\pm\rangle,$$

$$= \mp\frac{1}{2}(1 - \cos\vartheta(\mathbf{x}))\nabla_{\mathbf{x}}\varphi(\mathbf{x})$$

$$\mathscr{V}_\pm(\mathbf{x}) = \frac{1}{2m}\left|\langle\mp|\hat{U}^\dagger(\vartheta(\mathbf{x}), \varphi(\mathbf{x}))\nabla_{\mathbf{x}}\hat{U}(\vartheta(\mathbf{x}), \varphi(\mathbf{x}))|\pm\rangle\right|^2$$

$$= \frac{1}{8m}\left(|\nabla_{\mathbf{x}}\vartheta(\mathbf{x})|^2 + \sin^2\vartheta(\mathbf{x})|\nabla_{\mathbf{x}}\varphi(\mathbf{x})|^2\right) \equiv \mathscr{V}(\mathbf{x}), \qquad (2.89)$$

in addition to the APSs

$$\varepsilon_\pm(\mathbf{x}) = \pm\frac{1}{2}\mu b(\mathbf{x}), \qquad (2.90)$$

which define the gap function $\Delta(\mathbf{x}) = \mu b(\mathbf{x})$.

As a concrete example, we consider trapped atoms in a combination of a homogeneous magnetic field

$$\mathbf{b}_h = \left(\frac{b_0}{\sqrt{2}}, -\frac{b_0}{\sqrt{2}}, b_z \right) \tag{2.91}$$

and a magnetic field

$$\mathbf{b}_w(x', y') = \frac{\mu_0 I_w}{2\pi} \frac{1}{x'^2 + y'^2} (-y', x', 0) \tag{2.92}$$

produced by a thin wire, carrying an electrical current I_w in the z-direction, on a microfabricated chip (μ_0 is the magnetic constant). This kind of setup has been used to control atoms [15] and provides an excellent scenario for testing the Born–Oppenheimer approximation experimentally. By putting $\xi = \mu_0 I_w/(2\pi)$, the low-field seeking spin states are trapped in the vicinity of the point $x_0 = y_0 = \xi/(\sqrt{2}b_0)$, where the total magnetic field $\mathbf{b} \equiv \mathbf{b}_h + \mathbf{b}_w$ takes its minimum value $|b_z|$. By focusing on the motion close to (x_0, y_0), it is convenient to redefine coordinates $x' = x_0 + x$, $y' = y_0 + y$, and expand $\mathbf{b}$ to first order in x and y yielding

$$\mathbf{b}(x, y) \approx \left(\frac{b_0^2}{\xi} x, -\frac{b_0^2}{\xi} y, b_z \right). \tag{2.93}$$

By using cylindrical coordinates $(x, y, z) = (\rho \cos \phi, \rho \sin \phi, z)$, one finds

$$\varphi(\mathbf{x}) = -\phi, \quad \vartheta(\mathbf{x}) = \arccos \frac{b_z}{\sqrt{b_z^2 + \frac{b_0^4}{\xi^2}\rho^2}}, \tag{2.94}$$

which gives the potentials,

$$\mathbf{A}^{(+)}(\rho, \phi) = \frac{1}{2\rho} \left(1 - \frac{b_z}{\sqrt{b_z^2 + \frac{b_0^4}{\xi^2}\rho^2}} \right) \mathbf{e}_\phi \equiv A_\phi(\rho)\mathbf{e}_\phi,$$

$$\mathscr{V}(\rho) = \frac{1}{8m} \frac{b_0^4}{\xi^2} \frac{2b_z^2 + \frac{b_0^4}{\xi^2}\rho^2}{\left(b_z^2 + \frac{b_0^4}{\xi^2}\rho^2 \right)^2},$$

$$\varepsilon(\rho) = \frac{1}{2}\mu\sqrt{b_z^2 + \frac{b_0^4}{\xi^2}\rho^2} \tag{2.95}$$

for the low-field seeking spin states (we have assumed $\mu > 0$). We further obtain the synthetic magnetic field

$$
\mathbf{B}^{(+)} = \nabla_{\mathbf{x}} \times \mathbf{A}^{(+)}(\rho, \phi) = \frac{1}{\rho} \frac{\partial}{\partial \rho} \left(\rho A_\phi(\rho) \right) \mathbf{e}_z + \cdots
$$

$$
= \frac{b_z b_0^4}{2\xi^2} \frac{\mathbf{e}_z}{\left(b_z^2 + \frac{b_0^4}{\xi^2} \rho^2 \right)^{3/2}}. \tag{2.96}
$$

The scalar potential $\varepsilon(\rho) + \mathscr{V}(\rho)$ and synthetic magnetic field $B_z(\rho)$ are shown in Fig. 2.3 for a small but non-zero bias field b_z. The effect of the induced gauge potentials is essentially localised to the vicinity of the minimum of the APS $\varepsilon(\rho)$. Indeed, when the atom encircles the origin around a circular loop of radius ρ_0, it picks up a Berry phase

$$
\gamma(\mathscr{C}) = \oint_{\mathscr{C}} \mathbf{A}^{(+)}(\rho_0, \phi) \cdot d\mathbf{x} = \pi \left(1 - \frac{b_z}{\sqrt{b_z^2 + \frac{b_0^4}{\xi^2} \rho_0^2}} \right), \tag{2.97}
$$

which tends to a constant value far away from the origin. This shows that the model exhibits an approximate Aharonov–Bohm setting with a synthetic flux carrying half a flux unit and sitting at the origin [16]. This effect becomes exact in the $b_z \rightarrow 0$ limit, where the APS intersect with the lower surface and the induced gauge potentials take the form

$$
\mathbf{A} = \frac{1}{2\rho} \mathbf{e}_\phi,
$$

$$
\mathscr{V}(\rho) = \frac{1}{8m} \frac{1}{\rho^2}. \tag{2.98}
$$

The vector potential yields the Berry phase π for all non-zero radii, thus establishing the perfect Aharonov–Bohm analogy. The scalar potential defines a repulsive inverse-cubic force, known to be a generic feature of CIs [17]. This force has a stabilising effect as it repels the low-field seeking atoms from the origin, where they can undergo non-adiabatic transitions to the lower system and thereby leaving the trap.

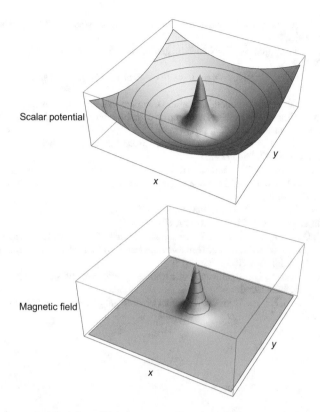

Fig. 2.3 Scalar potential $\varepsilon(\rho) + \mathscr{V}(\rho)$ (upper panel) and synthetic magnetic field $B_z(\rho)$ (lower panel) of the low-field seeking atoms for small but non-zero bias field b_z. The effect of the induced gauge potentials and field is localised to the vicinity of the minimum of the APS. The scalar gauge potential contributes with a repulsive force that prevents the atoms to approach the origin where they may undergo a non-adiabatic transition to the unstable lower energy surface (not shown). The synthetic magnetic field almost vanishes everywhere, except close to the origin, thus realising an approximate analogue of an Aharonov–Bohm system. This analogue becomes perfect in the $b_z \to 0$ limit

References

1. Messiah, A.: Quantum Mechanics, vol. II. North-Holland, Amsterdam (1962)
2. Tong, D.M.: Quantitative condition is necessary in guaranteeing the validity of the adiabatic approximation. Phys. Rev. Lett. **104**, 120401 (2010)
3. Marzlin, K.-P., Sanders, B.C.: Inconsistency in the application of the adiabatic theorem. Phys. Rev. Lett. **93**, 160408 (2004)
4. Ortigoso, J.: Quantum adiabatic theorem in light of the Marzlin-Sanders inconsistency. Phys. Rev. A **86**, 032121 (2012)
5. Roland, J., Cerf, N.J.: Quantum search by local adiabatic evolution. Phys. Rev. A **65**, 042308 (2002)
6. Wilczek, F., Zee, A.: Appearance of gauge structure in simple dynamical systems. Phys. Rev. Lett. **52**, 2111 (1984)

7. Berry, M.V.: Quantal phase factors accompanying adiabatic evolution. Proc. R. Soc. Lond. Ser. A **392**, 45 (1984)
8. Kult, D., Åberg, J., Sjöqvist, E.: Noncyclic geometric changes of quantum states. Phys. Rev. A **74**, 022106 (2006)
9. Provost, J.P., Vallee, G.: Riemannian structure on manifolds of quantum states. Commun. Math. Phys. **76**, 289 (1980)
10. Zanardi, P., Rasetti, M.: Holonomic quantum computation. Phys. Lett. A **264**, 94 (1999)
11. Duan, L.M., Cirac, J.I., Zoller, P.: Geometric manipulation of trapped ions for quantum computation. Science **292**, 1695 (2001)
12. Wu, T.T., Yang, C.N.: Concept of nonintegrable phase factors and global formulation of gauge fields. Phys. Rev. D **12**, 3845 (1975)
13. Mead, C.A., Truhlar, D.G.: On the determination of Born-Oppenheimer nuclear motion wave functions including complications due to conical intersections and identical nuclei. J. Chem. Phys. **70**, 2284 (1979)
14. Mead, C.A., Truhlar, D.G.: Conditions for the definition of a strictly diabatic electronic basis for molecular systems. J. Chem. Phys. **77**, 6090 (1982)
15. Cassettari, D., Chenet, A., Folman, R., Haase, A., Hessmo, B., Krüger, P., Maier, T., Schneider, S., Calarco, T., Schmiedmayer, J.: Micromanipulation of neutral atoms with nanofabricated structures. Appl. Phys. B **70**, 721 (2000)
16. Aharonov, Y., Bohm, D.: Significance of Electromagnetic Potentials in the Quantum Theory. Phys. Rev. **115**, 485 (1959)
17. Berry, M.V., Lim, R.: The Born-Oppenheimer electric gauge force is repulsive near degeneracies. J. Phys. A **23**, L655 (1990)

Chapter 3
Conical Intersections in Molecular Physics

Abstract The large mass difference between nuclei and electrons makes the Born–Oppenheimer approximation a central tool in the modelling of molecular systems. The mass difference permits an approximate separation of variables that provides insights into the structure and dynamics of molecules. Here, we discuss various aspects of conical intersections (CIs) between electronic APSs. We delineate the conditions for the appearance of CIs and discuss topological tests that can be used to detect intersections. We examine the role of symmetry for the existence of intersection points, in the particular case of Jahn–Teller systems. We discuss how CIs may influence the nuclear dynamics, causing subtle interference effects on nuclear wave packets, as well as physical effects on the vibrational spectrum that can be detected spectroscopically.

3.1   Introduction

A key problem in applied quantum theory is to solve the Schrödinger equation for a many-electron system interacting with a set of positively charged nuclei. This problem applies to molecules as well as to condensed matter systems. Here, we focus on the molecular case and postpone the discussion of condensed matter systems to Chap. 4.

© Springer Nature Switzerland AG 2020   33
J. Larson et al., *Conical Intersections in Physics*, Lecture Notes in Physics 965,
https://doi.org/10.1007/978-3-030-34882-3_3

The basic molecular Hamiltonian with $\mathcal{N}_n$ nuclei and $\mathcal{N}_e$ electrons with positions $\mathbf{Q} \equiv (\mathbf{Q}_1, \ldots \mathbf{Q}_{\mathcal{N}_n})$ and $\mathbf{q} \equiv (\mathbf{q}_1, \ldots \mathbf{q}_{\mathcal{N}_e})$, respectively, reads

$$
\begin{aligned}
\hat{H} &= \sum_{i_n=1}^{\mathcal{N}_n} \frac{\hat{\mathbf{P}}_{i_n}^2}{2m_{i_n}} + \sum_{i_e=1}^{\mathcal{N}_e} \frac{\hat{\mathbf{P}}_{i_e}^2}{2\mu_{i_e}} + \sum_{i_e < j_e}^{\mathcal{N}_e} \frac{e^2}{4\pi \epsilon_0} \frac{1}{\left| \hat{\mathbf{q}}_{i_e} - \hat{\mathbf{q}}_{j_e} \right|} \\
&\quad + \sum_{i_n < j_n}^{\mathcal{N}_n} \frac{e^2}{4\pi \epsilon_0} \frac{Z_{i_n} Z_{j_n}}{\left| \hat{\mathbf{Q}}_{i_n} - \hat{\mathbf{Q}}_{j_n} \right|} - \sum_{i_e=1}^{\mathcal{N}_e} \sum_{i_n=1}^{\mathcal{N}_n} \frac{e^2}{4\pi \epsilon_0} \frac{Z_{i_n}}{\left| \hat{\mathbf{q}}_{i_e} - \hat{\mathbf{Q}}_{i_n} \right|} \\
&\equiv \sum_{i_n=1}^{\mathcal{N}_n} \frac{\hat{\mathbf{P}}_{i_n}^2}{2m_{i_n}} + \hat{h}_e(\mathbf{Q}).
\end{aligned}
\tag{3.1}
$$

The electronic Hamiltonian $\hat{h}_e(\mathbf{Q})$ depends parametrically on the nuclear configuration $\mathbf{Q}$ and the eigenvalues $\varepsilon_n(\mathbf{Q})$ of $\hat{h}_e(\mathbf{Q})$ are electronic APSs. The problem of solving the Schrödinger equation for the Hamiltonian in (3.1) can be simplified by using that the nuclei are heavier and thereby move slower than the electrons; thus, the molecular energy eigenvalue problem can be addressed by using the Born–Oppenheimer approximation. In this picture, the nuclei move on a single APS $\varepsilon_n(\mathbf{Q})$ under influence of the synthetic gauge fields induced by the parametric dependence of the corresponding electronic eigenstate $|\psi_n(\mathbf{Q})\rangle$. The existence of non-trivial gauge fields depends crucially on the existence of conical intersections (CIs) in nuclear configuration space, which are therefore important to predict, detect, and characterise.

In this chapter, we focus on the appearance of synthetic gauge fields due to CIs in molecules. Section 3.2 describes different types of intersections between electronic APSs and how these intersections can be detected. The effects of the synthetic gauge fields have been detected experimentally in molecular spectroscopy. Section 3.3 discusses molecular systems undergoing Jahn–Teller distortions, which is the symmetry-breaking effect caused by nuclear vibrations interacting with degenerate electronic states in non-collinear molecules. As we will show, synthetic gauge fields may lead to non-trivial effects in Jahn–Teller molecules. Finally, we discuss the dynamical aspects of intersections between electronic APSs. As a first example, we show how wave packets describing the nuclear motion may show a destructive Aharonov–Bohm-like interference effect when encircling a synthetic flux line located at a CI. This interference effect has been proposed (but not verified experimentally yet) to be visible in scattering chemical reactions of small molecular systems, such as hydrogen exchange $H + H_2 \rightarrow H_2 + H$. In the second example, we demonstrate how the synthetic gauge field arising from the CI gives rise to dynamics reminiscent of a Hall effect.

3.2 Where Electronic Adiabatic Potential Surfaces Cross: Intersection Points

From the discussion in the previous chapter, we expect intersections between electronic APSs to play an important role for the synthetic gauge structure of molecular systems. We consider intersections that show up at isolated points, lines, etc., all defining manifolds with smaller dimension than that of the nuclear configuration space; thus, we exclude the Kramers degeneracy in the following. When approaching such a degenerate manifold, one would expect non-trivial synthetic gauge fields that may modify the vibrational spectra of the molecule, as well as cause subtle interference effects on the nuclear motion. Thus, it becomes important to characterise different types of intersections as well as to develop tools for finding them.

3.2.1 The Existence of Intersections

Let us for a moment ignore the kinetic energy of the nuclei and focus on the eigenvalue problem of the electronic Hamiltonian. We assume no external fields, which implies that the translational and rotational degrees of freedom decouple from the internal vibrational modes of the molecule. We focus on the vibrations and thereby view $\mathbf{Q}$ as a set of $\mathcal{N}_{vib}$ vibrational coordinates (*normal modes*) with $\mathcal{N}_{vib} = 3\mathcal{N}_n - 6$ and $3\mathcal{N}_n - 5$ for non-collinear and collinear molecules, respectively. We are interested in a situation where two of the electronic APSs coincide at a degenerate manifold $\mathcal{M}$, meaning that these surfaces coincide on $\mathcal{M}$. We therefore restrict the electronic Hamiltonian to just two electronic states $|1(\mathbf{Q})\rangle$ and $|2(\mathbf{Q})\rangle$ and consider the truncated electronic Hamiltonian

$$\hat{h}_e(\mathbf{Q}) = \sum_{k,l=1}^{2} h_{kl}(\mathbf{Q})|k(\mathbf{Q})\rangle\langle l(\mathbf{Q})|, \tag{3.2}$$

with $h_{kl}(\mathbf{Q})$ being a Hermitian 2×2 matrix. By solving the 2×2 eigenvalue problem, one sees that the resulting electronic APSs $\varepsilon_{\pm}(\mathbf{Q})$ cross at $\mathbf{Q}_0$ provided

$$h_{11}(\mathbf{Q}_0) - h_{22}(\mathbf{Q}_0) \equiv \Delta(\mathbf{Q}_0) = 0, \quad h_{12}(\mathbf{Q}_0) = 0. \tag{3.3}$$

Since $h_{12}(\mathbf{Q})$ is in general a complex-valued function of $\mathbf{Q}$, one needs to vary three vibrational coordinates in order to satisfy the conditions in (3.3). It follows that $\dim \mathcal{M} = \mathcal{N}_{vib} - 3$. For example, in a non-collinear triatomic molecule, two electronic APSs can cross only at isolated points in nuclear configuration space unless some additional symmetry is present, such as if $h_{12}(\mathbf{Q})$ is real-valued for all $\mathbf{Q}$ (which for instance applies to molecules with vanishing spin–orbit coupling).

The conditions $\Delta(\mathbf{Q}_0) = 0$, $\text{Re}[h_{12}(\mathbf{Q}_0)] = 0$, and $\text{Im}[h_{12}(\mathbf{Q}_0)] = 0$ define three hypersurfaces in nuclear configuration space. Figure 3.1 shows a typical shape of these hypersurfaces for a triatomic molecule with vibrational coordinates $\mathbf{Q} = (Q_x, Q_y, Q_z)$. For complex-valued $h_{12}(\mathbf{Q})$, the conditions for the two electronic APSs to cross is met at isolated points, while the crossing can occur along a line in parameter space if $\text{Im}[h_{12}(\mathbf{Q})]$ vanishes for all $\mathbf{Q}$. We further note that a non-crossing rule applies to the diatomic case, as a single nuclear coordinate is not sufficient to enforce all conditions in (3.3).

Having identified the conditions for when we may expect intersections, we now address the detailed behaviour of the electronic APSs when approaching $\mathcal{M}$. This

Fig. 3.1 The three conditions $\Delta(\mathbf{Q}) = 0$, $\text{Re}[h_{12}(\mathbf{Q})] = 0$, and $\text{Im}[h_{12}(\mathbf{Q})] = 0$ define three surfaces, where the last condition is coloured red (upper panel), for a triatomic molecule with vibrational coordinates $\mathbf{Q} = (Q_x, Q_y, Q_z)$. The corresponding APSs become degenerate at an isolated point in nuclear configuration space where all conditions are satisfied simultaneously. For a real-valued Hamiltonian, the last condition drops out and the APSs intersect along a line in nuclear configuration space (lower panel)

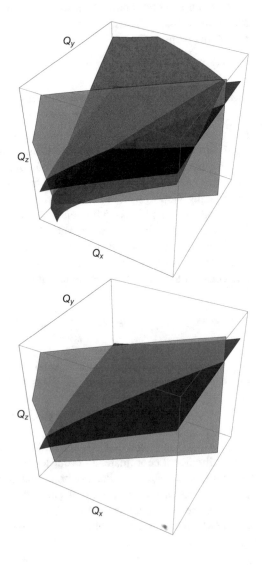

behaviour is determined by the derivatives of $\Delta(\mathbf{Q})$ and $h_{12}(\mathbf{Q})$, evaluated at $\mathbf{Q}_0 \in \mathscr{M}$, in a direction that lifts the degeneracy. To see this, we assume that $\mathbf{Q} = \mathbf{Q}_0 + (0,\ldots,\delta Q_v,\ldots,0) \notin \mathscr{M}$ for $\delta Q_v \neq 0$, and expand the APSs to second order yielding

$$
\varepsilon_\pm(\mathbf{Q}) = \varepsilon(\mathbf{Q}_0) + \delta Q_v \varepsilon'(\mathbf{Q}_0) + \frac{1}{2}\delta Q_v^2 \varepsilon''(\mathbf{Q}_0) \pm \frac{1}{2}|\delta Q_v| \sqrt{(\Delta'(\mathbf{Q}_0))^2 + |h'_{12}(\mathbf{Q}_0)|^2}
$$

$$
\pm \frac{1}{4}\delta Q_v |\delta Q_v| \frac{\Delta'(\mathbf{Q}_0)\Delta''(\mathbf{Q}_0) + 4\mathrm{Re}\left(h'_{12}(\mathbf{Q}_0)h''_{12}(\mathbf{Q}_0)^*\right)}{\sqrt{(\Delta'(\mathbf{Q}_0))^2 + |h'_{12}(\mathbf{Q}_0)|^2}}
$$

$$
+ O\left(\delta Q_v^3\right). \tag{3.4}
$$

Here, $\varepsilon(\mathbf{Q}) \equiv \frac{1}{2}(h_{11}(\mathbf{Q}) + h_{22}(\mathbf{Q}))$, $f'(\mathbf{Q}_0) = \partial_v f(\mathbf{Q})|_{\mathbf{Q}=\mathbf{Q}_0}$, $f''(\mathbf{Q}_0) = \partial_v^2 f(\mathbf{Q})|_{\mathbf{Q}=\mathbf{Q}_0}$, and we have used (3.3). Figure 3.2 shows a typical behaviour of the electronic APSs as a function of the displacement in the degeneracy-lifting coordinate Q_v. We see that $\varepsilon_\pm(\mathbf{Q})$ form a CI at $\delta Q_v = 0$.

The conical feature of the intersection originates from the linear displacement term proportional to $|\delta Q_v|$ in (3.4). If this term vanishes, the two electronic APSs typically form a *glancing intersection* at $\delta Q_v = 0$, see Fig. 3.3. We may note that the realisation of such a glancing intersection is highly demanding as it requires

$$
\Delta'(\mathbf{Q}_0) = 0, \quad h'_{12}(\mathbf{Q}_0) = 0, \tag{3.5}
$$

to be satisfied, on top of the conditions in (3.3). We thus expect that CIs are more likely than glancing intersections in polyatomic molecules. In fact, it has been shown that CIs are much more common than local minima of the energy gap, i.e., if one

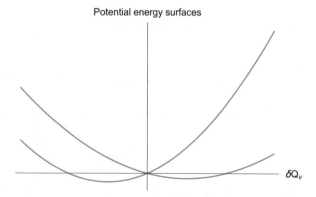

Fig. 3.2 Typical behaviour of two intersecting electronic APSs as a function of the displacement in the degeneracy-lifting coordinate Q_v. The two surfaces form a CI at $\delta Q_v = 0$

Potential energy surfaces

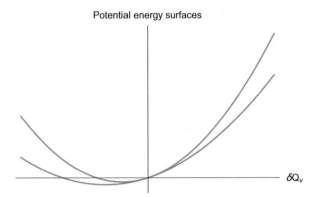

Fig. 3.3 Two electronic APSs forming a glancing intersection at $\delta Q_v = 0$

encounters a very small energy gap it is much more likely that one is close to a CI than an avoided intersection [1].

The number of conditions for intersecting electronic APSs depend on the degree of degeneracy. For real and symmetric Hamiltonian, one can prove that an r-fold degeneracy requires $\frac{1}{2}r(r + 1) - 1$ independent conditions [2]. Thus, one needs, for instance, to vary five vibrational coordinates in order to enforce a triple degeneracy, which implies at least four atoms in the molecule. The argument behind this rule is based on the observation that an r-fold degeneracy at $\mathbf{Q}_0$ means that Rank $(h_{kl}(\mathbf{Q}_0) - \varepsilon(\mathbf{Q}_0)\delta_{kl}) = N - r$, N being the number of electronic states used in the truncated electronic Hamiltonian. The restriction on the rank can be shown to be equivalent to $\frac{1}{2}r(r + 1)$ conditions for real and symmetric Hamiltonians. This number is reduced by one upon elimination of $\varepsilon(\mathbf{Q}_0)$.

3.2.2 Topological Tests for Intersections

The conditions for intersections between electronic APSs may be difficult to implement in practice as they require a detailed knowledge of $h_e(\mathbf{Q})$, which is typically a highly complex object in polyatomic molecules. This can, however, be circumvented by using topological techniques. In this way, existence criteria can be deduced, which can be used to search for intersections without detailed knowledge about $h_e(\mathbf{Q})$. Here, we discuss two such *topological tests* for intersections. These tests are complementary in the sense that one of them can be used when the other one fails and vice versa.

The tests apply to the case where the electronic Hamiltonian matrix $h_{kl}(\mathbf{Q}) = \langle k(\mathbf{Q})|\hat{h}_e(\mathbf{Q})|l(\mathbf{Q})\rangle$ can be put on real-valued form by an appropriate choice of electronic orthonormal basis $\{|k(\mathbf{Q})\rangle\}_{k=1}^{N}$, where all $\langle \mathbf{q}|k(\mathbf{Q})\rangle$ are real-valued functions. The eigenvectors of $h_{kl}(\mathbf{Q})$ can in this case always be chosen real; more precisely they can be expressed as linear combinations of the $\mathbf{Q}$-dependent basis vectors $|k(\mathbf{Q})\rangle$ with real-valued expansion coefficients $c_k(\mathbf{Q})$. Thus, they live on the $(N-1)$-dimensional sphere $S^{N-1} : \sum_{k=1}^{N} c_k^2(\mathbf{Q}) = 1$. We can first formulate the following theorem:

> If a real-valued electronic eigenvector changes sign around a loop $\mathscr{C}$ in nuclear configuration space, then there is at least one point on any surface bounded by $\mathscr{C}$ where it becomes degenerate with another energy eigenstate.

The proof is by *reductio ad absurdum*: By assuming the mapping from any surface bounded by $\mathscr{C}$ to S^{N-1} is smooth leads to a contradiction. This can be seen by noting that the process of dividing $\mathscr{C}$ into smaller and smaller pieces must eventually end up in an arbitrarily small loop around which the electronic eigenstate changes sign and thereby contradict the smoothness assumption. This topological test was designed by Longuet-Higgins [2] and has been used to find CIs in various molecular systems, e.g., triatomic systems such as LiNaK [3] and ozone [4].

By having established a sufficient criterion for intersecting electronic eigenstates, one may wonder whether this is also a necessary condition: If an electronic eigenvector does not change sign around $\mathscr{C}$, does this imply that there are no intersection points inside $\mathscr{C}$? To see that this is *not* the case, let us consider the electronic 3×3 Hamiltonian matrix

$$
h(Q_x, Q_y) = \begin{pmatrix} 0 & 0 & Q_y \\ 0 & 0 & Q_x \\ Q_y & Q_x & 0 \end{pmatrix}, \tag{3.6}
$$

which apparently has a triple degeneracy at the origin $(Q_x, Q_y) = (0, 0)$. Still, none of the electronic eigenvectors

$$
\psi_1(\phi) = \frac{1}{\sqrt{2}}(\sin\phi, \cos\phi, 1),
$$

$$
\psi_2(\phi) = (-\cos\phi, \sin\phi, 0)
$$

$$
\psi_3(\phi) = \frac{1}{\sqrt{2}}(\sin\phi, \cos\phi, -1) \tag{3.7}
$$

(we use polar coordinates $(Q_x, Q_y) = \rho(\cos\phi, \sin\phi)$) change sign when encircling the origin.

The failure of the Longuet-Higgins test to find some intersection points apparently raises the question if we can do better. Thus, we ask: Can we design other tests that can find intersection points even when the electronic eigenvectors do not change sign? We now discuss how this indeed can be done by using tools from algebraic topology. The idea is based on the observation that all loops in nuclear configuration space that can be deformed smoothly into each other form elements of the fundamental group Π_1, for which the zero element consists of all loops that can be contracted to a point.

Let us first apply Π_1 to a single electronic eigenvector. A well-known result is

$$\Pi_1\left(S^{N-1}\right) = \begin{cases} \mathbb{Z}, \text{ for } N = 2, \\ 0, \text{ for } N \geq 3. \end{cases} \tag{3.8}$$

Thus, the fundamental group of single vectors can be used to detect intersection points for $N = 2$, but fails to provide any non-trivial topological information for $N \geq 3$. In other words, a single vector is insufficient for $N \geq 3$.

To extend the applicability of the fundamental group, we instead use the observation that since none of the electronic eigenvectors changes sign around the loop $\mathscr{C}$, the set $\{\psi_k(\mathbf{Q}) = (c_1^k(\mathbf{Q}, \ldots, c_N^k(\mathbf{Q}))^T\}_{k=1}^N$ of electronic eigenstates can be taken to be a positively oriented orthonormal basis of $\mathbb{R}^N$. Every such basis can be viewed as an element of the special rotation group $SO(N)$ in dimension N. In other words, the $N \times N$ matrix $(\psi_1(\mathbf{Q}), \ldots, \psi_N(\mathbf{Q}))$ is an element of $SO(N)$. The difference between $SO(N)$ and S^{N-1} is that the former can be non-trivial also for $N \geq 3$. Indeed, it is known that

$$\Pi_1\left(SO(N)\right) = \begin{cases} \mathbb{Z}, \text{ for } N = 2, \\ \mathbb{Z}_2, \text{ for } N \geq 3, \end{cases} \tag{3.9}$$

the latter meaning that a loop in $SO(N \geq 3)$ is either trivial (can be contracted to a point) or is trivial only when traversed twice. We can now state the following theorem [5]:

> If the N eigenvectors of the electronic Hamiltonian matrix represent a non-trivial loop in $SO(N)$ when taken continuously around $\mathscr{C}$, then there must be at least one intersection point among the electronic eigenstates on every surface bounded by $\mathscr{C}$.

The proof is again by *reductio ad absurdum*, and follows exactly the same logic as in the proof of the Longuet-Higgins theorem above. This result makes it possible to detect the presence of a degeneracy by considering eigenvectors on a loop in nuclear configuration space even if they do not change sign around the loop. It can be shown [5] that this is an optimal test, in the sense that it is impossible to design a test that can do better in finding intersection points when the Longuet-Higgins test fails.

As an illustration, let us apply the test to the three-level model in (3.6). As pointed out above, this system has an intersection at the origin of the two-dimensional configuration space that cannot be detected by the Longuet-Higgins test since none of the eigenvectors in (3.7) changes sign. Now, the relevant rotation group for this model system is $SO(3)$, which is diffeomorphic to the real projective space RP^2, which can be viewed as the three-dimensional ball, here with radius π, where opposite points are identified. The points in the ball are the *rotation vectors* $\mathbf{v} \equiv \zeta\mathbf{n}$ representing the different rotations in $\mathbb{R}^3$. Here, $\zeta \in [-\pi, \pi]$ and $\mathbf{n}$ are the rotation angle and axis, respectively, where the absolute value $|\zeta|$ is defined as the length of the rotation vector, thus $\mathbf{n}$ is taken to be a unit vector. The identification of opposite points follows from the fact that the rotation vectors $\pm\pi\mathbf{n}$ describe the same rotation in $\mathbb{R}^3$. A non-trivial loop in SO(3) corresponds to a smooth sequence of rotations which pierce the surface of the ball and reappears at the opposite side. Now, by using the eigenvectors in (3.7), one finds the rotation axis and angle

$$\mathbf{n}(\phi) = \frac{1}{N(\phi)}\left(-\cos\phi, \sin\phi - 1, (1-\sqrt{2})\cos\phi\right),$$

$$\zeta(\phi) = \arccos\left(-1 + \frac{1}{2}\left(1 - \frac{1}{\sqrt{2}}\right)(1-\sin\phi)\right), \qquad (3.10)$$

$N(\phi)$ being a normalisation factor. The components of the rotation vector are shown in Fig. 3.4 for a loop around the origin. The change of sign at $\phi = \frac{\pi}{2}$ confirms the existence of the three-fold intersection point inside the loop.

The generalisation to complex-valued Hamiltonians was given by Stone [6]. Stones' test is based on the second homotopy group Π_2, basically looking at the Berry phase for continuous sequences of loops forming a sphere in nuclear configuration space. By requiring the Berry phase to vary continuously, a Berry phase change $\Delta\gamma$ over the sequence being a non-zero multiple of 2π signals an

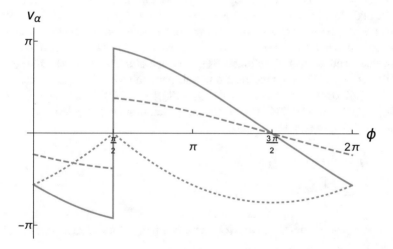

Fig. 3.4 Test for finding an intersecting point in the three-level model system in (3.6). We consider a loop around the three-fold intersection point at the origin of nuclear configuration space. The loop is parametrised by the polar angle $\phi \in [0, 2\pi]$. The Longuet-Higgins test fails to detect the intersection point since none of the electronic eigenvectors in (3.7) changes sign around the loop. On the other hand, the sign-preservation implies that the eigenvectors can be ordered so as to form a smoothly changing positively oriented basis around the loop so that they therefore define an SO(3) element for each ϕ. The components of the SO(3) rotation vector $\mathbf{v}(\phi) = \zeta(\phi)\mathbf{n}(\phi)$, defined by (3.10), are shown as function of ϕ with first, second, and third components given by the solid, dotted, and dashed lines, respectively. The change of sign of $\mathbf{v}(\phi)$ at $\phi = \frac{\pi}{2}$ shows that the loop is a non-trivial element of $\Pi_1(\text{SO}(3))$, which confirms the existence of the three-fold intersection point inside the loop

intersection point inside the sphere. Contrary to the Longuet-Higgins test for real-valued Hamiltonians, Stones' test is optimal and thereby exhausts all topological information about intersections of complex-valued electronic Hamiltonians, since $\Pi_2(SU(N))$ is non-trivial for all N [7].

3.2.3 The Molecular Aharonov–Bohm Effect on the Nuclear Motion

We now re-introduce the kinetic energy of the nuclei and examine the dynamical effects of the electronic intersection points. Since these points signal a breakdown of the single-surface picture associated with the Born–Oppenheimer approximation, great care is needed in order to deal with situations where the nuclear configuration is expected to move in their vicinity. Here, we discuss the *molecular Aharonov–Bohm effect* [8]. This effect is caused by a singular synthetic gauge field that influence the nuclei when they are confined to move on a single APS around a CI. We focus on the case where the electronic Hamiltonian can be taken to be real-valued for the relevant part of nuclear configuration space.

Let $\mathbf{Q} = (Q_x, Q_y)$ be the degeneracy-breaking nuclear displacements of the real-valued Hamiltonian. We assume mass-weighted coordinates defined by means of the transformation $Q_j \mapsto \sqrt{m_j} Q_j$ and restrict to a two-dimensional electronic subspace spanned by $|j(\mathbf{Q})\rangle$, $j = 1, 2$, where $\langle \mathbf{q}|j(\mathbf{Q})\rangle$ are assumed to be real-valued and smooth functions of $\mathbf{Q}$. The operator $\hat{P}(\mathbf{Q}) = |1(\mathbf{Q})\rangle\langle 1(\mathbf{Q})| + |2(\mathbf{Q})\rangle\langle 2(\mathbf{Q})|$ projects onto the electronic subspace and we define the corresponding 'local' Pauli operators $\hat{\sigma}_x(\mathbf{Q}) = |1(\mathbf{Q})\rangle\langle 2(\mathbf{Q})| + |2(\mathbf{Q})\rangle\langle 1(\mathbf{Q})|$ and $\hat{\sigma}_z(\mathbf{Q}) = |1(\mathbf{Q})\rangle\langle 1(\mathbf{Q})| - |2(\mathbf{Q})\rangle\langle 2(\mathbf{Q})|$. The molecular Hamiltonian projected onto the local electronic subspace reads ($\hbar = 1$)

$$
\begin{aligned}
\widetilde{\hat{H}} &= \hat{P}(\mathbf{Q})\hat{H}\hat{P}(\mathbf{Q}) \\
&= -\frac{1}{2}\hat{P}(\mathbf{Q})\partial_\alpha \partial^\alpha \hat{P}(\mathbf{Q}) + \varepsilon(\mathbf{Q})\hat{P}(\mathbf{Q}) \\
&\quad + \frac{1}{2}\sqrt{\Delta^2(\mathbf{Q}) + 4h_{12}^2(\mathbf{Q})}\left(\sin\varphi(\mathbf{Q})\hat{\sigma}_x(\mathbf{Q}) + \cos\varphi(\mathbf{Q})\hat{\sigma}_z(\mathbf{Q})\right), \quad (3.11)
\end{aligned}
$$

where we have defined $\tan\varphi(\mathbf{Q}) = h_{12}(\mathbf{Q})/\Delta(\mathbf{Q})$. In the Born–Oppenheimer approximation regime, we focus on the lower energy molecular state, which reads

$$
\begin{aligned}
|\widetilde{\Psi}(\mathbf{Q})\rangle &= |\psi_-(\mathbf{Q})\rangle\chi_-(\mathbf{Q}) \\
&= \left(-\sin\frac{\varphi(\mathbf{Q})}{2}|1(\mathbf{Q})\rangle + \cos\frac{\varphi(\mathbf{Q})}{2}|2(\mathbf{Q})\rangle\right)\chi_-(\mathbf{Q}). \quad (3.12)
\end{aligned}
$$

Note that the electronic wave function $\langle \mathbf{q}|\psi_-(\mathbf{Q})\rangle$ is real-valued, which implies that the corresponding Berry phase factor only can take the values ± 1.

We now consider a loop $\mathscr{C} : [0, 1] \ni s \mapsto \mathbf{Q}(s) | \mathbf{Q}(1) = \mathbf{Q}(0)$ and make a smooth phase choice such that $|j(\mathbf{Q}(1))\rangle = |j(\mathbf{Q}(0))\rangle$, $j = 1, 2$. The angle $\varphi(\mathbf{Q})$ is determined by the electronic part of the Hamiltonian in (3.11) up to an integer multiple of 2π. Thus,

$$\varphi(\mathbf{Q}(1)) = \varphi(\mathbf{Q}(0)) + n \times 2\pi, \quad n \text{ integer}, \tag{3.13}$$

which implies that

$$|\psi_-(\mathbf{Q}(1))\rangle = (-1)^n |\psi_-(\mathbf{Q}(0))\rangle. \tag{3.14}$$

In the case of odd n, the Longuet-Higgins theorem implies that there must exist a CI inside $\mathscr{C}$. Since the molecular state $|\Psi(\mathbf{Q})\rangle$ is a solution of the Schrödinger equation it must be single-valued also in the sign changing case, which can be assured by using the single-valued section

$$|\tilde{\psi}_-(\mathbf{Q})\rangle = e^{i\varphi(\mathbf{Q})/2} |\psi_-(\mathbf{Q})\rangle \tag{3.15}$$

resulting in the effective Schrödinger equation

$$-\frac{1}{2} \left(\partial_\alpha - \frac{i}{2} \partial_\alpha \varphi(\mathbf{Q}) \right) \left(\partial^\alpha - \frac{i}{2} \partial^\alpha \varphi(\mathbf{Q}) \right) \chi_-(\mathbf{Q}) + \left(\frac{1}{8} \partial_\alpha \varphi(\mathbf{Q}) \partial^\alpha \varphi(\mathbf{Q}) \right.$$

$$\left. + \varepsilon(\mathbf{Q}) - \frac{1}{2} \sqrt{\Delta^2(\mathbf{Q}) + 4h_{12}^2(\mathbf{Q})} \right) \chi_-(\mathbf{Q}) = \tilde{E} \chi_-(\mathbf{Q}) \tag{3.16}$$

for the single-valued nuclear 'wave function' $\chi_-(\mathbf{Q})$. By using the definition of the mixing angle $\varphi(\mathbf{Q})$, we find the synthetic vector potential

$$\mathbf{A}^\alpha = \frac{1}{2} \partial^\alpha \varphi(\mathbf{Q}) = \frac{1}{2} \left(\frac{\Delta(\mathbf{Q}) \big(\partial^\alpha h_{12}(\mathbf{Q}) \big) - \big(\partial^\alpha \Delta(\mathbf{Q}) \big) h_{12}(\mathbf{Q})}{h_{12}^2(\mathbf{Q}) + \Delta^2(\mathbf{Q})} \right), \tag{3.17}$$

which is singular where $h_{12}^2(\mathbf{Q}_0) + \Delta^2(\mathbf{Q}_0) = 0$. This coincides with the intersection points between the electronic states. By choosing nuclear coordinates so that a conical intersection occurs at $\mathbf{Q}_0 = \mathbf{0}$, and assuming that $h_{12}(\mathbf{Q}) \approx h'_{12}(\mathbf{0}) Q_x$ and $\Delta(\mathbf{Q}) \approx \Delta'(\mathbf{0}) Q_y$, we find

$$\mathbf{A}^\alpha(\mathbf{Q}) = \frac{1}{2} h'_{12}(\mathbf{0}) \Delta'(\mathbf{0}) \frac{(Q_y, -Q_x)}{\big(h'_{12}(\mathbf{0}) \big)^2 Q_x^2 + \big(\Delta'(\mathbf{0}) \big)^2 Q_y^2}. \tag{3.18}$$

To calculate the corresponding Berry phase for a loop $\mathscr{C}$ around the origin, we put $h'_{12}(\mathbf{0})Q_x = \rho \cos \phi$ and $\Delta'(\mathbf{0})Q_y = \rho \sin \phi$ yielding

$$\gamma(\mathscr{C}) = \oint_{\mathscr{C}} \mathbf{A}(\mathbf{Q}) \cdot d\mathbf{Q} = -\frac{1}{2} \oint_{\mathscr{C}} d\phi = -\pi, \qquad (3.19)$$

which explicitly unfolds the synthetic Aharonov–Bohm-like structure corresponding to half a flux unit, a π-*magnetic flux*, located at the intersection point.

The synthetic Aharonov–Bohm flux has a physical effect on the vibrational spectrum that has been experimentally verified in the metallic trimer Na_3 [9]. This effect shows up as a half-integral quantisation of the angular momentum related to the internal pseudo-rotational motion of the nuclei. Explicitly, the effect can be seen by assuming a harmonic localisation potential $\varepsilon(\mathbf{Q}) = \frac{1}{2}\omega^2 \rho^2$ and isotropic linear coupling ($h'_{12}(\mathbf{0}) = \Delta'(\mathbf{0})$), as well as considering large displacements and low lying states, yielding the approximate rovibrational spectrum

$$\widetilde{E}_{n,J} = \left(n + \frac{1}{2}\right)\omega + AJ^2, \qquad (3.20)$$

where $n = 0, 1, \ldots$, A is determined by the displacement amplitude, and, due to the synthetic flux, $|J| = \frac{1}{2}, \frac{3}{2}, \ldots$. The half-odd quantisation of J can be observed spectroscopically in the transitions between the energy levels displayed in (3.20).

3.3 The Jahn–Teller Effect

We have considered the existence of electronic intersections and how these influence the nuclear motion. However, up to this point we have not addressed the origin of these intersections. We shall here discuss the relation between symmetry and intersection points in molecules. Specifically we are concerned with the role played by the discrete point group symmetries of molecules to predict different types of intersections of electronic APSs.

3.3.1 Spontaneous Breaking of Molecular Symmetry: The Jahn–Teller Theorem

As argued before, one should expect CIs to be abundant in polyatomic molecules, while the glancing intersections should be considerably more rare, simply because they require additional constraints (cf., e.g., (3.5)). This observation implies that intersection points are typically not stable with respect to degeneracy-breaking deformations of a molecule. Surprisingly, though, this is not always true: if the degeneracy is induced by the axial symmetry of a collinear nuclear configuration (such as for carbon dioxide CO_2), then it will be stable to deformations out of its axial symmetry. On the other hand, if the degeneracy is associated with some dim ≥ 2 irreducible representation of a non-collinear molecule, then there is no symmetry reason why this degeneracy should be stable. These two results is the essence of the *Jahn–Teller theorem* [10] and the symmetry-lowering is known as the *Jahn–Teller effect* [11, 12]. The Jahn–Teller (JT) effect is known to occur, e.g., in octahedral complexes of transition metals, leading to elongated or compressed stable structures corresponding to the symmetry breaking $O_h \rightarrow D_4$, but also in small molecules such as the metallic trimers Li_3 and Na_3, where the JT effect induces the symmetry reduction $D_{3h} \rightarrow C_{2v}$.

We now outline the proof of the JT theorem. Let us recall that the electronic Hamiltonian is the sum of the kinetic energy of the electrons and the potential energy for a *fixed* nuclear configuration. Thus, the electronic Hamiltonian describes a set of negatively charged quantum particles moving in the field of a set of positive point-charges. If a configuration $\mathbf{Q}_0$ of these classical charges is invariant under a set of discrete transformations forming a point group, the corresponding electronic eigenstates must belong to an irreducible representation of this point group and the dimension of this irreducible representation coincides with the degree of degeneracy. To determine the stability of such a degeneracy, the proof considers the expansion of the potential energy $V(\mathbf{q}, \mathbf{Q})$ in symmetry adapted nuclear coordinates Q_ν around $\mathbf{Q}_0$. By differentiating the electronic eigenvalue equation, one finds the leading term

$$\delta Q_\nu \partial_\nu \varepsilon_n(\mathbf{Q})|_{\mathbf{Q}=\mathbf{Q}_0} = \delta Q_\nu \langle \psi_{n;k}^T(\mathbf{Q}_0)|\partial_\nu V(\mathbf{q}, \mathbf{Q})\big|_{\mathbf{Q}=\mathbf{Q}_0} |\psi_{n;k}^T(\mathbf{Q}_0)\rangle, \quad (3.21)$$

where $k = 1, \ldots, \dim T$, T being the irreducible representation of the electronic eigenstate. If Q_ν belongs to the identity representation then there is no reason of symmetry that the right-hand side of (3.21) should vanish. However, the value of a symmetrical coordinate can be varied at will without destroying the symmetry of the molecule so that the symmetry-induced degeneracy remains. By finding a value of the symmetrical coordinate for which $\partial_\nu \varepsilon_n(\mathbf{Q})|_{\mathbf{Q}=\mathbf{Q}_0}$ would determine the stable size of the system, just as the minimum of the ground state energy potential curve of a diatomic molecule determines the equilibrium distance between the atoms. For a symmetry-breaking coordinate belonging to an irreducible representation T', on the other hand, the right-hand side of (3.21) would only vanish by symmetry if the product $T \times T'$ does not contain T. By going through all possible molecular point groups, it turns out that $T \times T'$ *always* contains T for non-collinear nuclear configurations, while it *never* contains T for collinear nuclear configurations.

In the following, we discuss in detail the $E \times \epsilon$ model, which is arguably the 'canonical' JT system that comprises the main features of spontaneous symmetry-breaking and molecular Aharonov–Bohm effect. We focus on the non-trivial synthetic gauge fields associated with the CIs arising when including a combination of linear and quadratic coupling terms in this model system.

3.3.2 The $E \times \epsilon$ JT Model

The $E \times \epsilon$ JT model system describes a doubly degenerate electronic subspace (E) at some high-symmetry point $\mathbf{Q}_0$ coupled to a pair (ϵ) of symmetry-lowering vibrational normal modes (Q_x, Q_y). These vibrations are degenerate, i.e., associated with the same vibrational frequency ω. We assume the electronic states belong to the E irreducible representation of the equilateral triangle configuration of some triatomic homonuclear molecule, such as Na_3 or Li_3. In this case, the degenerate vibrational mode together with the symmetric mode Q_a, which determines the 'size' of the molecule, is a complete set of vibrational coordinates.

One may picture a continuous variation of the symmetry-breaking nuclear coordinates as a path in the 'space of shapes' of the molecule. It is convenient to use polar coordinates ρ, ϕ defined as $(Q_x, Q_y) = \rho(\cos\phi, \sin\phi)$ to picture the path. Here, ρ measures the size of the symmetry-breaking deformation, while ϕ describes an internal rotation, a *pseudo-rotation*, in the space of molecular shapes. In order to study the synthetic gauge structure of the model, it becomes essential to examine the Berry phase for pseudo-rotational loops in nuclear configuration space, i.e., for paths where ϕ varies over non-zero multiples of 2π.

By setting up the model, we fix the electronic states at the symmetry point and restrict to the two-dimensional electronic subspace $\mathrm{Span}\{|j\rangle \equiv |j(\mathbf{Q}_0)\rangle\}$, $j = 1, 2$. This enforces a diabatic, or more precisely *crude adiabatic*, approach to the system description, in which the nuclear kinetic energy becomes diagonal. By expanding around $\mathbf{Q}_0$, one finds to second order in the symmetry-breaking coordinates [13]

$$\tilde{H} = -\frac{1}{2}\nabla^2 + \frac{1}{2}\rho^2 + \left(k\rho\cos\phi + \frac{1}{2}g\rho^2\cos 2\phi \right)\sigma_z$$
$$+ \left(k\rho\sin\phi - \frac{1}{2}g\rho^2\sin 2\phi \right)\sigma_x, \qquad (3.22)$$

where we have set $\omega = \hbar = 1$ for notational simplicity, and σ_x, σ_z are the standard Pauli operators on $\mathrm{Span}\{|j\rangle\}$. Here, k and g are the real-valued linear and quadratic coupling strengths, respectively, between the nuclear and electronic degrees of freedom. The Jahn–Teller theorem entails that k is typically non-zero, which implies that the lower APS has its minimum at $\rho \neq 0$. By diagonalising $h_e(\rho, \phi)$ we find the electronic APSs

$$\varepsilon_\pm(\rho, \phi) = \frac{1}{2}\rho^2 \pm \sqrt{k^2\rho^2 + kg\rho^3\cos 3\phi + \frac{1}{4}g^2\rho^4} \qquad (3.23)$$

with the mixing angle

$$\varphi(\rho, \phi) = \arctan\left(\frac{k\sin\phi - \frac{1}{2}g\rho\sin 2\phi}{k\cos\phi + \frac{1}{2}g\rho\cos 2\phi} \right). \qquad (3.24)$$

We note that these solutions satisfy a three-fold symmetry in the pseudo-rotational angle ϕ. In particular, one can identify four CIs, one at $\rho = 0$ and three at $\rho = 2k/g$ distributed at pseudo-rotational angles $\phi = \pi/3, \pi$, and $5\pi/3$. These CIs all collapse to a single glancing intersection at $\rho = 0$ in the strong quadratic limit (the Renner–Teller model) where $k/g \to 0$ and the three outer CIs are thrown to infinity in the strong linear coupling limit where $k/g \to \infty$, leaving a single CI at the origin and the lower APS takes a 'sombrero'-like shape. In the following, we examine the synthetic gauge structure associated with these four CIs.

In the Born–Oppenheimer approximation regime, the Hamiltonian contains synthetic scalar and vector potential terms arising from the parameter dependence of the mixing angle φ. One finds,

$$\mathscr{V}(\rho, \phi) = \frac{1}{8} |\nabla \varphi(\rho, \phi)|^2$$

$$= \frac{1}{8\rho^2} \frac{\frac{1}{4}k^2 g^2 \rho^2 \sin^2 3\phi + \left(k^2 - \frac{1}{8}g^2\rho^2 - \frac{1}{2}kg\rho \cos 3\phi\right)^2}{\left(k^2 + \frac{1}{4}g^2\rho^2 + kg\rho \cos 3\phi\right)^2}.$$

$$\mathbf{A}(\rho, \phi) = \frac{1}{2}\nabla\varphi(\rho, \phi) = -\frac{1}{4} \frac{kg \sin 3\phi}{k^2 + \frac{1}{8}g^2\rho^2 + kg\rho \cos 3\phi} \mathbf{e}_\rho$$

$$+ \frac{1}{2\rho} \frac{k^2 - \frac{1}{8}g^2\rho^2 - \frac{1}{2}kg\rho \cos 3\phi}{k^2 + \frac{1}{4}g^2\rho^2 + kg\rho \cos 3\phi} \mathbf{e}_\phi. \tag{3.25}$$

The lowest electronic APS $\varepsilon_- + \mathscr{V}$ is shown in Fig. 3.5. The three-fold pseudo-rotational symmetry can be clearly seen. A strong repulsive force originating from the synthetic part of the APS creates a high barrier for the nuclei to reach the conical intersections where the Born–Oppenheimer approximation breaks down.

Figure 3.6 shows Berry phase $\gamma(\rho)$ for counterclockwise circular paths of radius ρ. We identify two regions for which the Berry phase is constant and separated by the radius $\rho = 2k/g$ where it is undefined. Explicitly, $\gamma(0 < \rho < 2k/g) = \pi$ since the path encircles only the CI at the origin over this region, while all four CIs are encircled when $\rho > 2k/g$, which results in a jump of -2π since the three CIs at $\rho = 2k/g$ are each associated with a Berry phase of $-\pi$. The sign difference can be understood from the orientation of the flow lines around the four CIs, as shown in Fig. 3.7. Note that the Berry phase factor $e^{i\gamma(\rho > 2k/g)}$ is trivial, while the non-zero underlying Berry phase $\gamma(\rho > 2k/g)$ entails the existence of intersection point(s) inside the loops, in accordance with the generalised test outlined in Eqs. (3.8) and (3.9).

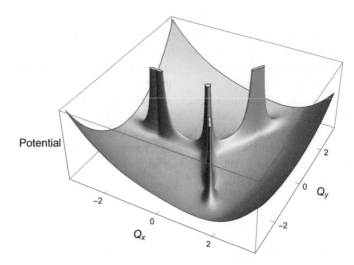

Fig. 3.5 Lowest electronic APS as a function of the symmetry-breaking vibrational coordinates Q_x and Q_y. The linear and quadratic coupling strengths are chosen to be $k = g = 0.1$. The nuclear motion is repelled from the CIs at the origin as well as at $\rho = 2k/g = 2$ for $\phi = \pi/3, \pi, 5\pi/3$, by a strong repulsive force associated with scalar gauge potential $\mathcal{V}(\rho, \phi)$

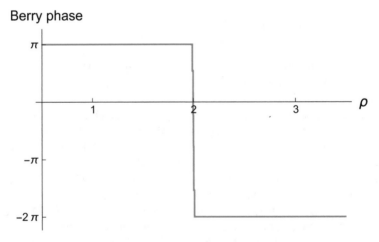

Fig. 3.6 Berry phase for circular counterclockwise paths of radius ρ. The linear and quadratic coupling strengths are chosen to be $k = g = 0.1$

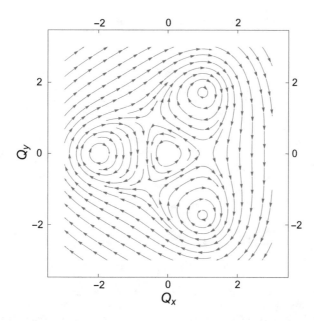

Fig. 3.7 Flow lines of the synthetic vector potential. The linear and quadratic coupling strengths are chosen to be $k = g = 0.1$. The flow is singular at the CIs; it has a counterclockwise orientation for the CI at the origin and a clockwise orientation for the three CIs at $\rho = 2\,k/g = 2$

3.4 Dynamical Manifestation of Conical Intersections

JT models are simple examples that show how geometric aspects, such as the Berry phase, arise. In the gauge formalism of these models we saw that we can envision a magnetic flux penetrating the CI. This fact has led to that people talk about a molecular Aharonov–Bohm effect, as described above in Sect. 3.2.3. Encircling the CI should somehow manifest in the dynamics.

In [14], Schön and Köppel considered a 'self-interference' situation where an initial Gaussian wave-packet resides at the sombrero minima of the lower APS. As the wave-packet is free to expand it will predominantly spread along the sombrero minima. After a certain time, the initial wave-packet has spread around the full minimum and interferes with itself. The interference pattern will capture the π-magnetic flux through the CI. In Fig. 3.8 (left column) we show the wave-packet at three different times; shortly after it has been released, when it has spread out over roughly half the sombrero minima and finally when it has spread out completely. The model is the $E \times \epsilon$ JT model in (3.22) discussed in Sect. 3.3. The initial wave-packet is a Gaussian with a width corresponding to the ground state of the JT harmonic oscillator. To see the result of the magnetic Aharonov–Bohm flux we should compare the results one would obtain when there is no flux. Such a model is obtained from simply disregarding the upper APS and the non-adiabatic couplings between the different electronic eigenstates. The results of the corresponding

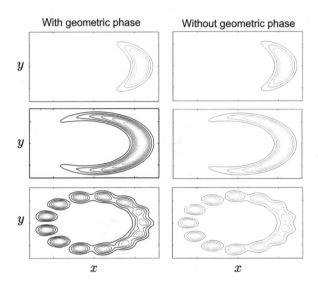

Fig. 3.8 Snapshots of how an initial Gaussian wave-packet, residing on the minima of the lower APS and with averages $(x, y) = (x_0, 0)$ and $(p_x, p_y) = (0, 0)$, evolves in the $E \times \epsilon$ JT model. The left plots displays the results of the full JT model, where we have a synthetic magnetic flux through the CI, while the plots to the right give the same when we have removed the upper APS (i.e., there is no synthetic magnetic flux). Before the self-interference of the wave-packet there is no evident differences between the two cases. However, as soon as the wave-packet interferes with itself we see that the interference pattern is shifted due to the Aharonov–Bohm flux; the constructive–destructive interferences have been interchanged

simulation are presented in the right column of the same figure. When the self-interference is manifested, the Aharonov–Bohm flux has the effect of shifting the interference pattern such that where one has constructive interference one instead finds destructive interference. Such a swapping of interference peaks/minima is expected since the wave-packet encircles a π-magnetic flux.

By looking at the wave-packet in Fig. 3.8, it seems as if it only knows about the CI, and the corresponding magnetic flux, once the interference is well developed. For the upper two rows there are no noticeable differences between the two cases. A natural question that comes to mind is therefore: is it only possible to catch the physics of the CI, and in particular the underlying gauge structure, in some interference setup? The answer is no.

Away from the CI a localised wave-packet should not be affected by the synthetic magnetic flux through the CI. If initialised on one of the two APSs with a vanishing average momentum its evolution for short times should be governed by the gradient of the potential, i.e., it should start to 'slide down' the potential. Upon approaching the CI, the Born–Oppenheimer approximation will break down and the different adiabatic states will start to mix. Let us assume that initially the wave-packet resides on the lower APS with an average $(\langle x \rangle, \langle y \rangle)_{t=0} = (x_0, 0)$. The symmetry of the problem suggests that $\langle y \rangle_t = 0$ for all times. The numerical simulation for such a situation is presented in Fig. 3.9a, which shows the time evolved trajectories $(\langle x \rangle, \langle y \rangle)_t$ for the $E \times \epsilon$ JT model. We see that the above discussion is wrong, the wave-packet bends and $\langle y \rangle_t \neq 0$ whenever $t > 0$. How can we understand such anomalous dynamics?

Mathematically, position and momentum are just labels of the variables, what we call what is unimportant. We can write the $E \times \epsilon$ JT model Hamiltonian as

$$\hat{H}_{E \times \epsilon} = \omega \left[\frac{(\hat{x} - \hat{A}_x)^2}{2} + \frac{(\hat{y} - \hat{A}_y)^2}{2} + \frac{p_x^2}{2} + \frac{p_y^2}{2} \right] + \hat{\Phi}, \qquad (3.26)$$

where $\hat{A}_{x,y} = -\frac{g}{\omega} \hat{\sigma}_{x,y}$ and $\hat{\Phi} = \frac{g^2}{\omega}$. Put in this form we may interpret $\hat{A}_{x,y}$ as components of the vector potential and $\hat{\Phi}$ as a scalar potential. In doing so we should think of x and y as the 'momentum' and p_x and p_y as the 'position'. In this picture, the particle moves in an isotropic two-dimensional harmonic potential and exposed to a synthetic magnetic field. The magnetic field is now non-Abelian since $[\hat{A}_x, \hat{A}_y] = 2i \left(\frac{g}{\omega} \right)^2 \hat{\sigma}_z \neq 0$. This results in a synthetic magnetic field and Lorentz force

$$\hat{\mathbf{B}}_z = -i[\hat{A}_x, \hat{A}_y] = 2 \left(\frac{g}{\omega} \right)^2 \hat{\sigma}_z, \qquad \mathbf{F}_{\text{Lor}} = 2 \left(\frac{g}{\omega} \right)^2 \hat{\sigma}_z (-\hat{y}, \hat{x}). \qquad (3.27)$$

Thus, the particle experiences a constant magnetic field in either positive or negative z-direction depending on its internal state. This is the reason why we see the bending of the trajectory in Fig. 3.9a. Later in Sects. 4.2.2 and 6.1.2 we will meet the same anomalous dynamics but in very different settings. The phenomenon just described is a manifestation of the *intrinsic spin Hall effect* (see Sect. 4.2.2) in the vibrational motion of molecules.

The linear $E \times \epsilon$ JT model is an idealisation of more realistic molecular models. In particular, the higher order corrections, quadratic and so on, may not be negligible. Furthermore, while CI's are not uncommon in general, to find them in the lower APS close to the vibrational ground state is rare. Li$_3$ is a molecule where the two lowest APSs supports a JT-like CI. The quadratic corrections are, however, non-vanishing and as a result one finds three additional CIs for non-zero positions. As long as the wave-packet evolution is confined to a region closer to the origin than the outer three CIs we may expect to see the spin Hall effect found in the linear $E \times \epsilon$ JT model. A numerical simulation of the trajectory $(\langle \hat{Q}_x \rangle, \langle \hat{Q}_y \rangle)_t$ for Li$_3$ is shown in Fig. 3.9b [15]. The APSs used for the simulation are obtained from *ab initio* quantum chemical calculations. The coordinates $Q_{x,y}$ are the vibrational normal modes of the molecule. The trajectory is only shown for short times since the wave-packet rapidly spreads and it becomes improper to talk about a trajectory for later times. Nevertheless, the transverse Hall motion is evident. Together with

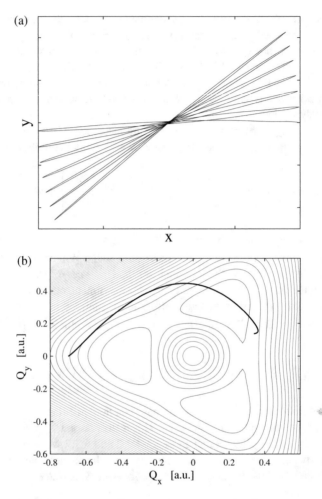

Fig. 3.9 The upper plot (**a**) gives a trajectory of the averages $(\langle \hat{x} \rangle, \langle \hat{y} \rangle)_t$ for an initial Gaussian wave-packet with means $(x, y) = (x_0, 0)$ and $(p_x, p_y) = (0, 0)$ for the linear $E \times \epsilon$ JT model. Naively one would expect that the polar symmetry of the model should imply that $\langle \hat{y} \rangle_t = 0$. Instead one finds that the wave-packet experiences a transverse force as it starts to 'slide down' the potential. In a 'dual' picture where we swap position and momentum, this force is identified with a synthetic Lorentz force and the anomalous motion is in fact a realisation of an intrinsic Hall effect. In the lower plot (**b**) we show the same but for a real physical system, the Li$_3$ molecule. Together with the trajectory we also plot the lower APS of Li$_3$ (obtained from *ab initio* quantum chemistry calculations). Here, the coordinates Q_x and Q_y are normal modes, and we use atomic units. Due to the strong JT coupling g of Li$_3$, the synthetic Lorentz force deflects the wave-packet strongly already during one oscillation, in fact in this example almost the whole wave-packet passes the CI on one side. The strong anharmonicity of the APSs implies that the initially localised Gaussian wave-packet rapidly spreads and the trajectory character is lost

the trajectory we also plot the lower APS for Li$_3$, and we can see how the quadratic corrections break the polar symmetry of the linear $E \times \epsilon$ JT model; the sombrero minimum splits up into three minima.

References

1. Truhlar, D.G., Mead, C.A.: Relative likelihood of encountering conical intersections and avoided intersections on the potential energy surfaces of polyatomic molecules. Phys. Rev. A **68**, 032501 (2003)
2. Longuet-Higgings, H.C.: The intersection of potential energy surfaces in polyatomic molecules. Proc. R. Soc. Lond. Ser. A **344**, 147 (1975)
3. Varandas, A.J.C., Tennyson, J., Murrell, J.N.: Chercher le croisement. Chem. Phys. Lett. **61**, 431 (1979)
4. Ceotto, M., Gianturco, F.A.: Charge-transfer effects in the gas-phase protonation of ozone: Locating the conical intersections. J. Chem. Phys. **112**, 5820 (2000)
5. Johansson, N., Sjöqvist, E.: Optimal topological test for degeneracies of real Hamiltonians. Phys. Rev. Lett. **92**, 060406 (2004)
6. Stone, A.J.: Spin-orbit coupling and the intersection of potential energy surfaces in polyatomic molecules. Proc. R. Soc. Lond. Ser. A **351**, 141 (1976)
7. Johansson, N., Sjöqvist, E.: Searching for degeneracies of real Hamiltonians using homotopy classification of loops in SO(n). Phys. Rev. A **71**, 012106 (2005)
8. Mead, C.A.: The molecular Aharonov-Bohm effect in bound states. Chem. Phys. **49**, 23 (1980)
9. von Busch, H., Dev, V., Eckel, H.-A., Kasahara, S., Wang, J., Demtröder, W., Sebald, P., Meyer, W.: Unambiguous proof for Berry's Phase in the sodium trimer: analysis of the transition $A^2 E'' \rightarrow X^2 E'$. Phys. Rev. Lett. **81**, 4584 (1998)
10. Jahn, H. A., Teller, E.: Stability of polyatomic molecules in degenerate electronic states. I. Orbital degeneracy. Proc. R. Soc. A. **161** 220 (1937)
11. Engleman, R.: The Jahn-Teller Effect in Molecules and Crystals. Wiley-Interscience, New York (1972)
12. Bersuker, I.: The Jahn-Teller effect. Cambridge University Press, Cambridge (2009)
13. Zwanziger, J.W., Grant, E.R.: Topological phase in molecular bound states: application to the $E \otimes e$ system. J. Chem.Phys. **87**, 2954 (1987)
14. Schön, J., Köppel, H.: Geometric phase effects and wave packet dynamics on intersecting potential energy surfaces. J. Chem. Phys. **103**, 9292 (1995)
15. Larson, J., Ghassemi, E.N., Larson, Å.: Anomalous molecular dynamics in the vicinity of a conical intersection. Europhys. Lett. **101**, 43001 (2013)

Chapter 4
Conical Intersections in Condensed Matter Physics

Abstract So far we have seen how conical intersections (CIs) appear in molecular systems, and how they may influence physical properties. These CIs showed up in the Born–Oppenheimer treatment and could be analysed in terms of synthetic gauge fields. Even though the adiabatic potential surfaces (APSs) often are given in normal vibrational coordinates, they still represent real space coordinates, i.e., the CIs appear in the real space. In this chapter, we see how CIs can occur in momentum space instead as so called *Dirac cones* or, as we will call them, *Dirac CIs*. The paradigmatic example of such Dirac CIs is graphene where carbon atoms form a two-dimensional hexagonal lattice. The periodicity of the lattice implies that the spectrum consists of energy bands restricted to the first Brillouin zone. Dirac CIs are point degeneracies of pairs of energy bands. We show, analogously to what we saw in the previous chapter, that there is a gauge structure connected to the physics of the bands. From this, one can define topological invariances like the *Chern number*. We discuss how the topological features are manifest in physical observables such as conductivity and edge states. While much of the chapter is devoted to periodic systems, we also give two examples where Dirac CIs emerge in continuum systems: spin–orbit coupled systems and superconductors. Superconductors differ from the other examples since the Dirac CIs appear at the mean-field level and require interaction between the electrons.

4.1 Band Theory

A classical particle in a periodic potential will either be trapped in one of the potential minima, or be free to move in the potential. In the first case, the energy of the particle is lower than the potential barriers that separate the minima, while in the second case its energy is larger than the barriers. In quantum mechanics, even with an energy below the barriers, the particle is still not limited to a single minimum since it can tunnel through the barrier to nearby minima. Thus, the particle is not bound, and the spectrum will not be discrete. However, as we will see, the spectrum is not simply connected; continuous energy bands are separated by energy gaps.

© Springer Nature Switzerland AG 2020

J. Larson et al., *Conical Intersections in Physics*, Lecture Notes in Physics 965,
https://doi.org/10.1007/978-3-030-34882-3_4

Clearly, the appearance of such windows of forbidden energies is a quantum effect, and is of crucial importance for material science.

In this section, we summarise some basics of band theory, starting with the important Bloch's theorem. The tight-binding model is introduced as the simplest example of how energy bands emerge in periodic Hamiltonians. Tight-binding models are often derived by expansion in a Wannier basis, which are localised states contrary to the extended Bloch states. We introduce the Bloch Hamiltonians, which are the momentum representation of the lattice model. In the Wannier basis, the Bloch Hamiltonian is readily diagonalisable, and it is the natural representation in which to discuss symmetries, i.e., time-reversal, particle-hole, and chiral symmetries. The presence or absence of these discrete symmetries determines possible topological properties of the model as discussed in the end of this section.

4.1.1 Bloch's Theorem

We assume a periodic potential, such that $V(\mathbf{x}) = V(\mathbf{x} + \mathbf{R_n})$ with the *Bravais vector* $\mathbf{R_n} = n_1\mathbf{e}_1 + n_2\mathbf{e}_2 + n_3\mathbf{e}_3$, where $\mathbf{e}_i$ are unit vectors along the edges of a unit cell and n_i integers. The set $\{\mathbf{R_n}\}$ defines a lattice in $\mathbb{R}^3$. The translation operator for a displacement $\mathbf{R_n}$ is $\hat{\mathscr{T}}(\mathbf{R_n}) = \exp\left(-i\hat{\mathbf{p}}\cdot\mathbf{R_n}/\hbar\right)$, which clearly must commute with the Hamiltonian $\left[\hat{\mathscr{T}}(\mathbf{R_n}), \hat{H}\right] = 0$. Note also that any two translational operators mutually commute. This together implies that we can always find a common basis for $\hat{H}$ and $\hat{\mathscr{T}}(\mathbf{R_n})$. Moreover, due to the unitarity of the translation operator and the relation

$$\hat{\mathscr{T}}(\mathbf{R_n} + \mathbf{R}_m) = \hat{\mathscr{T}}(\mathbf{R_n})\hat{\mathscr{T}}(\mathbf{R}_m) \tag{4.1}$$

we conclude that the eigenvalues of $\hat{\mathscr{T}}(\mathbf{R_n})$ are of the form $c(\mathbf{R_n}) = \exp(-i\mathbf{k}\cdot\mathbf{R}_n)$ for a real parameter $\mathbf{k}$ called *Bloch vector*. Let us consider an eigenstate $\Psi(\mathbf{x})$ of the translational operator, i.e., $\hat{\mathscr{T}}(\mathbf{R_n})\Psi(\mathbf{x}) = \exp(-i\mathbf{k}\cdot\mathbf{R}_n)\Psi(\mathbf{x})$. From the definition of the translation operator as displacing $\mathbf{x}$ by $\mathbf{R_n}$ we have

$$\Psi(\mathbf{x} + \mathbf{R}_n) = e^{-i\mathbf{k}\cdot\mathbf{R}_n}\Psi(\mathbf{x}). \tag{4.2}$$

This is Bloch's theorem; the energy eigenstates of a periodic Hamiltonian are periodic in each Bravais vector up to an overall phase.

Usually the eigenstates, or *Bloch waves*, are written as

$$\Psi(\mathbf{x}) = e^{-i\mathbf{k}\cdot\mathbf{x}}u_{\mathbf{k}}(\mathbf{x}), \tag{4.3}$$

introducing the *Bloch function* $u_{\mathbf{k}}(\mathbf{x})$ being periodic on the lattice, i.e., $u_{\mathbf{k}}(\mathbf{x} + \mathbf{R_n}) = u_{\mathbf{k}}(\mathbf{x}), \forall\mathbf{R_n}$. Note that $\Psi(\mathbf{x})$ is defined up to a *reciprocal lattice vector* $\mathbf{K}$, i.e., $\mathbf{k} \to \mathbf{k} + \mathbf{K}$. The reciprocal lattice is the Fourier transform of the original Bravais lattice, such that $\mathbf{K}$ is a vector in the reciprocal lattice. The largest volume around the origin where no two vectors can be connected by a reciprocal lattice vector is called the (first) *Brillouin zone*. It is thereby enough to consider vectors within the Brillouin zone, and hence remove the ambiguity of the Bloch vector for the state (4.2). When limiting the Bloch states to the first Brillouin zone, the Bloch vector $\mathbf{k}$ is then often referred to as *crystal-* or *quasi-momentum*. For a cubic Bravais lattice, with lattice constant a, the reciprocal lattice is also cubic with the lattice constant $2\pi/a$. Similarly, for a hexagonal lattice, with lattice constant a, the reciprocal lattice is hexagonal with the lattice constant $2\pi/a$. These are both examples of *self-dual* lattices, but not all lattices are self-dual.

4.1.2 Tight-Binding Model

In the next section, we continue discussing properties of the Bloch states and also introduce an alternative set of states, the Wannier states, but for now we make a short break and demonstrate how the energy band structure emerges in periodic potentials. We have already seen that we can assign a quantum number $\mathbf{k}$ to any Bloch state (4.3), but there is an additional quantum number labelling the corresponding energy band. To show this we consider a simple one-dimensional periodic lattice potential $V(x)$ as schematically depicted in Fig. 4.1.

In the limit of infinitely deep lattice potential wells, a particle at a given site n is bounded to that site, i.e., the tunnelling probability, or amplitude, is zero. Thus, we obtain an infinite set of decoupled potential wells, and we can denote the energy eigenstates by $|n, \nu\rangle$ and E_ν the corresponding energies. Here, n denotes the site index, and $\nu = 1, 2, \ldots$ the onsite energy eigenvalues. With the notations used in atomic physics, $|n, 1\rangle$ represents an s-orbital localised to site n, $|n, 2\rangle$ a p-orbital, and so on. It is clear that the ground state (and any other state) of the full Hamiltonian is infinitely degenerate; all states $|n, 1\rangle$ ($n \in \mathbb{Z}$) are degenerate ground states with energy E_1. As soon as there is a non-vanishing tunnelling amplitude t between the sites, these states are no longer energy eigenstates.

Let us assume a weak tunnelling amplitude between neighbouring sites. Hence, we neglect tunnelling between sites beyond nearest neighbours, which is essentially the *tight-binding approximation* (TBA). The deeper the lattice, the more accurate is the tight-binding approximation. We further assume that the tunnelling between sites preserves the index ν—other processes do not preserve energy and are greatly reduced. This is the so called *single-band approximation*. With these two approximations, and assuming the particle to reside in the lower states, we have the single-band tight-binding Hamiltonian

$$\hat{H}_{\text{TBA}} = E_1 \sum_{n=-\infty}^{+\infty} |n\rangle\langle n| - t \sum_{n=-\infty}^{+\infty} (|n\rangle\langle n+1| + |n+1\rangle\langle n|), \tag{4.4}$$

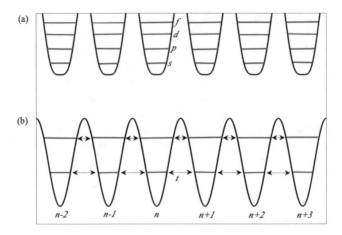

Fig. 4.1 Idea of the tight-binding model. The solid black lines represent the periodic lattice potential $V(x)$, and the red horizontal lines give a schematic picture of the onsite energies, corresponding to the orbital states s, p, d and so on. In (**a**) we envision a very deep lattice potential such that we can neglect tunnelling between neighbouring sites n and $n \pm 1$. When the lattice amplitude is decreased, lower plot (**b**), a particle at site n may tunnel to its neighbours $n \pm 1$. This tunnelling is marked by the double arrows, and the tunnelling amplitude between two s-orbitals is denoted t. Note that the tunnelling amplitude depends on the orbital states; the deeply lying s-orbitals have a lower tunnelling rate than the higher p-orbitals for example. As soon as t is non-zero, the set of degenerate onsite energies split and form a continuous 'band' of allowed energies, thus we form a band for each onsite orbital: s, p, d,...

where we implemented the usual convention of defining the tunnelling term with a minus sign. The Hamiltonian is diagonalised via a Fourier transform

$$|q\rangle = \sum_{n=-\infty}^{+\infty} e^{inq} |n\rangle, \tag{4.5}$$

and it is straightforward to check that this is indeed an energy eigenstate

$$\hat{H}_{\mathrm{TBA}}|q\rangle = E_1 \sum_{n=-\infty}^{+\infty} e^{inq} |n\rangle - t \sum_{n=-\infty}^{+\infty} (e^{inq} |n - 1\rangle + e^{inq} |n + 1\rangle)$$

$$\tag{4.6}$$

$$= [E_1 - 2t \cos(q)] \sum_{n=-\infty}^{+\infty} e^{inq} |n\rangle = [E_1 - 2t \cos(q)] |q\rangle.$$

Due to the tunnelling, the degenerate ground state energy E_1 has now split up in a band of energies ranging from $E_1 - 2t$ to $E_1 + 2t$.

In accordance with Bloch's theorem, the energy eigenstate $|q\rangle$ is also an eigenstate of the translational operator $\hat{\mathcal{T}}(a)$:

$$\hat{\mathcal{T}}(a)|q\rangle = \sum_{n=-\infty}^{+\infty} e^{inq}|n+1\rangle = e^{iq} \sum_{n=-\infty}^{+\infty} e^{inq}|n\rangle = e^{iq}|q\rangle, \qquad (4.7)$$

with a, as before, the lattice constant. Knowing that $|q\rangle$ is a Bloch state, from (4.2) we can identify $q = ka$. Thus, the eigenenergies, or energy dispersion, are $\epsilon(k) = E_1 - 2t\cos(ka)$ and we recall that the quasi-momentum is restricted to the first Brillouin zone $k \in [-\pi/a, +\pi/a)$.

What we have learnt is that for deep lattices, where we can assume that tunnelling takes place only between neighbouring sites, the infinite set of degenerate energies split to form an energy band centred at E_n and the width of the band is determined from the tunnelling amplitude t. We therefore can label the different bands with the ν-index, which in the atomic terminology becomes s-, p-, d-bands, and so on. As the tunnelling amplitude gets larger for higher onsite orbital states, the band width also increases with the band index ν. Including a non-zero tunnelling to next nearest neighbours alters the energy dispersions to become $\epsilon(k) = E_1 - 2t\cos(ka) - 2u\cos(2ka)$, where u is the next nearest neighbour tunnelling amplitude. Of course, in real materials the lattice structure can be rather complicated with several atoms per unit cell, and the full band spectrum can become way more complex.

4.1.3 Bloch and Wannier Functions

As we have seen, the energy eigenstate of a periodic Hamiltonian is labelled by the quasi-momentum $\mathbf{k}$ and the band index ν, i.e.,

$$\Psi_{\mathbf{k}\nu}(\mathbf{x}) = e^{-i\mathbf{k}\cdot\mathbf{x}}u_{\mathbf{k}\nu}(\mathbf{x}). \qquad (4.8)$$

These Bloch waves, which are extended over the full lattice, are orthogonal in both quantum numbers

$$\int d\mathbf{x}\, \Psi_{\mathbf{k}\nu}^*(\mathbf{x})\Psi_{\mathbf{k}'\nu'}(\mathbf{x}) = \delta_{\nu\nu'}\delta(\mathbf{k} - \mathbf{k}'), \qquad (4.9)$$

and a general state of the particle can be expanded in terms of the Bloch states,

$$\Psi(\mathbf{x}) = \int_{\mathbf{k}\in BZ} d\mathbf{k} \sum_{\nu=1}^{\infty} a_\nu(\mathbf{k})\Psi_{\mathbf{k}\nu}(\mathbf{x}). \qquad (4.10)$$

Here, BZ denotes the Brillouin zone. In the single-band approximation we restrict the second sum to a single term. We understand the different terms in the expansion as generalised plane waves with wave vectors $\mathbf{k}$. The periodic function $u_{\mathbf{k}\nu}(\mathbf{x})$ modulates the amplitude of the plane waves. In the previous section, we also introduced another basis, the site localised states $|n, \nu\rangle$, which in many circumstances can be more practical to work with. Such *Wannier states* can be formed from the Bloch states as a sort of Fourier transform, i.e., by integrating the quasi-momentum over the first Brillouin zone,

$$w_{\nu \mathbf{R_n}}(\mathbf{x}) = \frac{1}{V_{BZ}} \int_{BZ} d\mathbf{k}\, e^{i\mathbf{k}\cdot(\mathbf{x}-\mathbf{R_n})} u_{\nu \mathbf{k}}(\mathbf{x}), \tag{4.11}$$

where V_{BZ} is the Brillouin zone volume and $\mathbf{R_n}$ is the position of site $\mathbf{n}$. The Wannier functions form a complete orthonormal basis,

$$\int d\mathbf{x}\, w^*_{\nu \mathbf{R_n}}(\mathbf{x}) w_{\nu' \mathbf{R_{n'}}}(\mathbf{x}) = \delta_{\nu\nu'} \delta_{\mathbf{nn'}} \tag{4.12}$$

and how localised they are around $\mathbf{R_n}$ depends on the form of the Bloch states $u_{\nu \mathbf{k}}(\mathbf{x})$. In Fig. 4.2, we show the Wannier functions for the two lowest bands of the one-dimensional *Mathieu Hamiltonian*

$$\frac{H}{E_R} = -\frac{d^2}{dx^2} + V \cos^2(x), \tag{4.13}$$

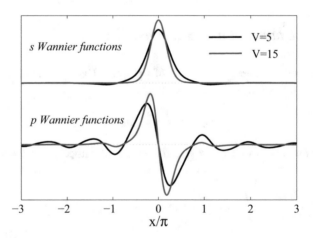

Fig. 4.2 Two examples of the Wannier functions for the s- and p-band for a potential $V \cos^2(x)$ (in dimensionless units). The upper curves give the Wannier function for the lowest band, and the lower curves give the Wannier functions for the first excited band. The orthogonality of the Wannier functions implies that they have to shift sign (the dotted thin line marks the zero line). This is clear for the p-band Wannier functions, but can also be seen for the s-band Wannier function; for the smaller potential amplitude (black curve) the Wannier function is seen to become negative at points. Note how the widths increase as the amplitude V of the potential is lowered

where $E_R = \hbar^2 k^2 / 2m$ is the *recoil energy* for the particle with mass m and $k = 2\pi/\lambda$ is the wave number. The orthogonality condition (4.12) implies that the Wannier functions must shift signs, as demonstrated in the figure. Note how the width of the Wannier functions grows with the decreased amplitude V of the potential.

It turns out that there is no unique way to construct site localised orthogonal states, i.e., (4.11) is just one example. Finding the transformation that results in the most localised states is an old problem and is of great importance for material science. For example, in Sect. 4.2 we will see how the orbital shape, and in particular hybridisation of different atomic orbitals, determine the type of bonding in crystals. Furthermore, the type of orbitals also directly influence the conductance properties of the particles in the lattice. The ambiguity in constructing Wannier functions derives from the fact that the Bloch states $u_{\nu\mathbf{k}}(\mathbf{x})$ are only determined up to an overall phase $\varphi(\mathbf{k})$. The choice of phase affects the shape of the resulting Wannier function (4.11). Kohn proved that for one-dimensional systems there exists a unique construction that results in exponentially localised Wannier functions [1]. In higher dimensions the question of exponentially localised Wannier functions is much more delicate. One general result is that exponentially localised Wannier functions exist if and only if all the Chern numbers (see (4.45) below) are zero [2]. Thus, the topology of the bands is reflected in the properties of the Wannier functions, and in particular for time-reversal symmetric systems, exponentially localised Wannier functions can always be constructed.

4.1.4 Single Particle Lattice Models and Bloch Hamiltonians

In Appendix A.1, we introduce the idea of second quantisation, which becomes an especially powerful tool when considering interacting many-body systems. Nothing prevents us, though, from using the formalism also for non-interacting systems. This might even be advantageous in order to make the step smaller when we turn to interacting systems. Starting from a second quantised single particle Hamiltonian, in this section we introduce the *Bloch Hamiltonians* which naturally emerge when one considers translationally invariant quadratic lattice models.

When working with a periodic model it is often convenient to expand the particle creation operator in the Wannier basis as

$$\hat{\psi}^\dagger(\mathbf{x}) = \sum_\nu \sum_{\mathbf{i}} \hat{c}^\dagger_{\nu\mathbf{i}} w_{\nu\mathbf{R_i}}(\mathbf{x}), \tag{4.14}$$

where $\hat{c}^\dagger_{\nu\mathbf{i}}$ creates a particle at site $\mathbf{i}$ and orbital state ν, i.e., $\hat{c}^\dagger_{\nu\mathbf{i}}|0\rangle = |1_{\nu\mathbf{i}}\rangle$. Similarly, the annihilation operators $\hat{\psi}(\mathbf{x})$ and $\hat{c}_{\nu\mathbf{i}}$ destroy a particle at either position $\mathbf{x}$ or at site $\mathbf{i}$. Here, we are limiting the analysis to a single particle, i.e., the total number operator $\hat{N} = \sum_\nu \sum_{\mathbf{i}} \hat{n}_{\nu\mathbf{i}} = 1$ with $\hat{n}_{\nu\mathbf{i}} = \hat{c}^\dagger_{\nu\mathbf{i}} \hat{c}_{\nu\mathbf{i}}$. The index ν could specify arbitrary internal degrees of freedom or it could label the corresponding band. Often it specifies the spin of the particle. At the single particle level, the statistics, whether the particle is a fermion, $\{\hat{c}_{vi}, \hat{c}_{\mu j}^{\dagger}\} = \delta_{v\mu}\delta_{ij}$, or a boson, $[\hat{c}_{vi}, \hat{c}_{\mu j}^{\dagger}] = \delta_{v\mu}\delta_{ij}$, plays no role. The second quantised Hamiltonian is

$$\hat{H} = \int d\mathbf{x}\, \hat{\Psi}^{\dagger}(\mathbf{x})\,\hat{H}_{1st}\hat{\Psi}(\mathbf{x}), \tag{4.15}$$

where $\hat{H}_{1st}$ is the first quantised Hamiltonian. The Hamiltonian clearly preserves the particle number $[\hat{N}, \hat{H}] = 0$, and moreover it is quadratic in the creation and annihilation operators $\hat{c}_{vi}^{\dagger}$, $\hat{c}_{vi}$,

$$\hat{H} = \sum_{v,\mu} \sum_{\mathbf{i},\mathbf{j}} \hat{c}_{vi}^{\dagger} h_{\mathbf{ij}}^{v\mu} \hat{c}_{\mu j}, \tag{4.16}$$

with the matrix elements given by

$$h_{\mathbf{ij}}^{v\mu} = \int d\mathbf{x}\, w_{v\mathbf{R_i}}^{*}(\mathbf{x})\,\hat{H}_{1st}w_{\mu\mathbf{R_j}}(\mathbf{x}). \tag{4.17}$$

In the tight-binding approximation, we restrict the second sum to only nearest neighbours; $\sum_{\mathbf{i},\mathbf{j}} \rightarrow \sum_{\langle \mathbf{ij}\rangle}$, and we assume the onsite energies $h_{\mathbf{ii}}^{v\mu}$ to be constant, and omitted them from the Hamiltonian. In doing so we assume that the onsite energies are the same for the two orbitals v and μ, e.g., their respective Wannier functions have the same shapes or we simply include only one internal state. Since we consider translationally invariant models we have no spatial dependences and we can write $h_{\mathbf{ij}}^{v\mu} = h^{v\mu}$ for the nearest neighbours.

If, for example, we consider a spin-1/2 particle confined to a single band we may write $v = \uparrow, \downarrow$ (given some basis) and by introducing

$$\hat{c}_{\mathbf{i}}^{\dagger} = \begin{bmatrix} \hat{c}_{\uparrow\mathbf{i}}^{\dagger} & \hat{c}_{\downarrow\mathbf{i}}^{\dagger} \end{bmatrix}, \qquad \hat{c}_{\mathbf{i}} = \begin{bmatrix} \hat{c}_{\uparrow\mathbf{i}} \\ \hat{c}_{\downarrow\mathbf{i}} \end{bmatrix}, \tag{4.18}$$

the Hamiltonian is written as

$$\hat{H} = \sum_{\langle \mathbf{ij}\rangle} \hat{c}_{\mathbf{i}}^{\dagger} \begin{bmatrix} h^{\uparrow\uparrow} & h^{\uparrow\downarrow} \\ h^{\downarrow\uparrow} & h^{\downarrow\downarrow} \end{bmatrix} \hat{c}_{\mathbf{j}}. \tag{4.19}$$

After Fourier transforming

$$\hat{c}_{vi} = \frac{1}{\sqrt{L}} \sum_{\mathbf{k}} \hat{c}_{v\mathbf{k}} e^{i\mathbf{k}\cdot\mathbf{i}}, \tag{4.20}$$

with L the number of sites, the Hamiltonian is diagonal in the momentum quantum number,

$$\hat{H} = \sum_{\mathbf{k}} \hat{c}_{\mathbf{k}}^{\dagger} h(\mathbf{k}) \hat{c}_{\mathbf{k}} = \sum_{\mathbf{k}} \hat{c}_{\mathbf{k}}^{\dagger} \begin{bmatrix} h^{\uparrow\uparrow}(\mathbf{k}) & h^{\uparrow\downarrow}(\mathbf{k}) \\ h^{\downarrow\uparrow}(\mathbf{k}) & h^{\downarrow\downarrow}(\mathbf{k}) \end{bmatrix} \hat{c}_{\mathbf{k}}, \tag{4.21}$$

where $h^{\nu\nu'}(\mathbf{k}) = h^{\nu\nu'}\left[\cos(k_x) + \cos(k_y) + \cos(k_z)\right]$. Here, it should be understood that the sum runs over the vector components of $\mathbf{k}$. To diagonalise the Hamiltonian is straightforward and gives the dispersions $\epsilon_{\pm}(\mathbf{k}) = \lambda_{\pm}\left[\cos(k_x) + \cos(k_y) + \cos(k_z)\right]$ and $\lambda_{\pm} = \frac{1}{2}\left(h^{\uparrow\uparrow} + h^{\downarrow\downarrow}\right) \pm \frac{1}{2}\sqrt{\left(h^{\uparrow\uparrow} - h^{\downarrow\downarrow}\right)^2 + 4|h^{\uparrow\downarrow}|^2}$.

In (4.21), $h(\mathbf{k})$ is the *Bloch Hamiltonian*. In this simple example, the particle number is conserved. For a quadratic Hamiltonian, the solvability does not rely on the particle conservation, and this symmetry can indeed be broken in many instances. A well-known example is the mean-field BCS theory of superconductors, which will be considered in Sect. 4.3.

When the unit cell contains more than a single atom, we need the same number of operators to describe the second quantised Hamiltonian. The dimension of the Bloch Hamiltonian may then be much larger than in the examples given above that described two bands. We will see an example of this in Sect. 4.4, where we discuss the spectrum of graphene.

4.1.5 Symmetries

In elementary courses in quantum mechanics, symmetries are usually associated with conserved quantitates described by unitary operators that commute with the Hamiltonian, i.e., $\hat{H} = \hat{U}\hat{H}\hat{U}^{\dagger}$. In the previous section, we mentioned particle conservation following from $[\hat{N}, \hat{H}] = 0$. The corresponding unitary would be $\hat{U}_N(\phi) = \exp\left(-i\hat{N}\phi\right)$ which clearly commutes with the Hamiltonian for any ϕ. This defines a continuous symmetry. The consequence is that we can group the states according to their number of particles, i.e., eigenstates of $\hat{N}$. The Hamiltonian then takes a block form within this grouping of states, each block can be assigned one particle number. In this section, we are interested in discrete symmetries that are not represented by some unitary invariance. We will discuss the three main ones, *time-reversal*, *chiral*, and *particle-hole symmetry*. The first is represented by an anti-unitary operator, which according to *Wigner's theorem* is acceptable as a physical transformation in the sense that it preserves the normalisation. The other two, however, are neither represented by unitary nor anti-unitary operators, but by operators that transforms the spectrum of the system. In other words, they tell something about the symmetry of the spectrum, and as such they do not represent proper symmetries which we are used to connect to conserved quantities (even though they have been labelled as 'symmetries' in the community).

4.1.5.1 Time-Reversal Symmetry

Time-reversal symmetry is the most important of the discrete symmetries. It should reflect the transformation $t \rightarrow -t$. Since time is not an observable that is characterised by an operator, there is no simple unitary reversing time. Instead one must consider the meaning of time-reversal on other observables. Any time-reversal operator $\hat{T}$ should flip the velocity of a particle, or $\hat{T}\hat{p}\hat{T}^{-1} = -\hat{p}$. Simultaneously, it should leave $\hat{x}$ unaltered, and we note

$$\hat{T}[\hat{x}, \hat{p}]\hat{T}^{-1} = -i \quad \Rightarrow \quad \hat{T}i\hat{T}^{-1} = -i. \tag{4.22}$$

Thus, the operator $\hat{T}$ is proportional to complex conjugation, which we denote by $\hat{K}$. In general, we can have

$$\hat{T} = \hat{V}\hat{K} \tag{4.23}$$

for some unitary operator $\hat{V}$. For a time-reversal symmetric system we have that its Hamiltonian commutes with $\hat{T}$, $[\hat{T}, \hat{H}] = 0$. To see that $\hat{T}$ is different from the unitary symmetries we study the action of $\hat{T}$ on physical states. For states transformed under $\hat{T}$, i.e., $|\tilde{\phi}\rangle = \hat{T}|\phi\rangle$, we have $\langle\tilde{\phi}|\tilde{\varphi}\rangle = \langle\phi|\varphi\rangle^*$ which tells us that $\hat{T}$ is *anti-unitary*. We find another property of the time-reversal operator by noting that the application of it twice must return the same state up to an overall phase,

$$\hat{T}^2 = \hat{V}\hat{K}\hat{V}\hat{K} = \hat{V}\hat{V}^* = \hat{V}(\hat{V}^T)^{-1} = e^{i\theta}\hat{1}. \tag{4.24}$$

Taking the transpose of the last identity, we find two equations

$$\hat{V} = e^{i\theta}\hat{V}^T, \qquad \hat{V}^T = \hat{V}e^{i\theta}, \tag{4.25}$$

which combined give

$$\hat{V} = e^{2i\theta}\hat{V}. \tag{4.26}$$

Thus, $e^{i\theta} = \pm 1$ or $\hat{T}^2 = \pm 1$. When $\hat{T}$ squares to -1 we have $\hat{V} = -\hat{V}^{-1}$. For spinless particles it is sufficient to choose $\hat{T} = \hat{K}$, and we have $\hat{T}^2 = 1$. Thus, in order to have the other situation where $\hat{T}^2 = -\hat{1}$, we must somehow involve spin.

Spin is an internal angular momentum and should transform under time-reversal as

$$\hat{T}\hat{\mathbf{S}}\hat{T}^{-1} = -\hat{\mathbf{S}}. \tag{4.27}$$

This means that the spin flips sign under time-reversal, which is equivalent to a π rotation around an arbitrary direction. The convention has been to choose the rotation around the y-axis. This defines the unitary operator of the time-reversal operator, i.e.,

$$\hat{T} = e^{-i\pi \hat{S}_y} \hat{K}. \tag{4.28}$$

We thus have

$$\hat{T}^2 = e^{-i\pi \hat{S}_y} \hat{K} e^{-i\pi \hat{S}_y} \hat{K} = e^{-i\pi \hat{S}_y} e^{i\pi \hat{S}_y^*} = e^{-i2\pi \hat{S}_y}. \tag{4.29}$$

Hence, acting twice with the time-reversal operator is the same as a 2π-rotation of the spin, or

$$\hat{T}^2 = \begin{cases} +\hat{1}, & \text{integer spin,} \\ -\hat{1}, & \text{half--odd integer spin.} \end{cases} \tag{4.30}$$

This is in agreement with that for spinless (i.e., spin-0) particles $\hat{T}^2 = \hat{1}$.

Assume that the Hamiltonian $\hat{H}$ is time-reversal invariant, and let $|\psi_n\rangle$ and E_n be an energy eigenstate and eigenvalue, respectively, then we have

$$\hat{H}|\psi_n\rangle = E_n|\psi_n\rangle \quad \Rightarrow \quad \hat{H}|\tilde{\psi}_n\rangle = E_n|\tilde{\psi}_n\rangle, \tag{4.31}$$

where, as above, $|\tilde{\psi}_n\rangle = \hat{T}|\psi_n\rangle$. In other words, $|\psi_n\rangle$ and $|\tilde{\psi}_n\rangle$ have the same energy E_n. If these two states are the same we have that they differ by at most a phase factor; $|\tilde{\psi}_n\rangle = \hat{T}|\psi_n\rangle = e^{i\delta}|\psi_n\rangle$ for some phase δ. However, if we act once more with $\hat{T}$ we get $\hat{T}^2|\psi_n\rangle = e^{-i\delta}\hat{T}|\psi_n\rangle = e^{-i\delta}e^{i\delta}|\psi_n\rangle = |\psi_n\rangle$, but this is a contradiction if the time-reversal operator squares to $-\hat{1}$ and we conclude that $|\psi_n\rangle$ and $|\tilde{\psi}_n\rangle$ are distinct physical states. This is the result of *Kramer's theorem*, stating that for half-integer spin systems that are time-reversal invariant the spectrum is at least doubly degenerate.

We end the discussion about time-reversal symmetry by looking at another spectral consequence of $\hat{T}$-invariance. The onsite annihilation operator $\hat{c}_{v\mathbf{i}}$ should be invariant under time-reversal, i.e., $\hat{T}\hat{c}_{v\mathbf{i}}\hat{T}^{-1} = \hat{c}_{v\mathbf{i}}$ provided we consider spinless particles. In Fourier space

$$\hat{T}\hat{c}_{v\mathbf{i}}\hat{T}^{-1} = \frac{1}{\sqrt{N}}\sum_{\mathbf{k}} e^{-i\mathbf{k}\cdot\mathbf{R}_{\mathbf{i}}}\hat{T}\hat{c}_{v\mathbf{k}}\hat{T}^{-1} = \frac{1}{\sqrt{N}}\sum_{\mathbf{k}} e^{-i\mathbf{k}\cdot\mathbf{R}_{\mathbf{i}}}\hat{c}_{v-\mathbf{k}}, \tag{4.32}$$

so that $\hat{T}\hat{c}_{v\mathbf{k}}\hat{T}^{-1} = \hat{c}_{v-\mathbf{k}}$. Assuming that the Hamiltonian is time-reversal invariant and given by (4.21), we find

$$\hat{T}\hat{H}\hat{T}^{-1} = \sum_{\mathbf{k}} \hat{c}^\dagger_{-\mathbf{k}}\hat{T}h(\mathbf{k})\hat{T}^{-1}\hat{c}_{-\mathbf{k}} = \hat{H} = \sum_{\mathbf{k}} \hat{c}^\dagger_{\mathbf{k}}h(\mathbf{k})\hat{c}_{\mathbf{k}}, \tag{4.33}$$

which implies that the Bloch Hamiltonian obeys

$$\hat{T}h(\mathbf{k})\hat{T}^{-1} = h(-\mathbf{k}).$$ (4.34)

Thus, for a system obeying time-reversal symmetry, the dispersions $E(\mathbf{k})$ are invariant under the parity shift $\mathbf{k} \leftrightarrow -\mathbf{k}$. The result is also true for spinful systems [3]. For half-integer spins, the two states corresponding to $\mathbf{k}$ and $-\mathbf{k}$ are the Kramer's pair of states. But what about points where $\mathbf{k} = -\mathbf{k}$ (up to any reciprocal lattice vector)? According to Kramer's theorem we must have a degeneracy at such point, implying that we have two Bloch states $\Psi_{\mathbf{k}v}(\mathbf{x})$ with same quasi-momentum and energy. As we will see, these are points where the bands may form Dirac cones, the reciprocal-space-counterpart of conical intersections.

4.1.5.2 Particle-Hole Symmetry

Particle-hole symmetry (also known as charge conjugation) originates from particle physics, where it defines a symmetry between particles and their antiparticles. For an electron, for example, it flips the sign of the charge, $e \rightarrow -e$. But this is similar to the relation between absence of electrons, i.e., holes, in the valence band and electrons in the conductance band. It is not a true symmetry as it is valid in some energy range around the Fermi surface. More precisely, in condensed matter physics many of the physical properties can be understood by exploring effective models that describe low energy excitations around the Fermi surface. We then look at a limited energy regime around the Fermi energy; excitations create a particle above the Fermi surface and a hole below it. Particle-hole symmetry becomes relevant in these scenarios. To explain it we consider a particular example, namely the Bogoliubov–de Gennes Bloch Hamiltonian (see Sect. 4.3)

$$h_{\mathrm{BdG}}(\mathbf{k}) = \begin{bmatrix} \epsilon(\mathbf{k}) & -\Delta_{\mathbf{k}}^* \\ -\Delta_{\mathbf{k}} & -\epsilon(\mathbf{k}) \end{bmatrix}$$ (4.35)

that describes the excitations of the superconductor. Its eigenvalues are given by $\lambda_{\pm}(\mathbf{k}) = \pm\sqrt{|\Delta_{\mathbf{k}}|^2 + \epsilon^2(\mathbf{k})}$, with the positive one connected with particle excitations and the negative with hole excitations.

Let us define the *charge conjugation* operator $\hat{C} = i\hat{\tau}_y\hat{K}$, where $\hat{\tau}_y$ is the Pauli operator acting on the particle/hole space, and $\hat{K}$ is as before complex conjugation. We note that

$$\hat{C}h_{\mathrm{BdG}}(\mathbf{k})\hat{C}^{-1} = \hat{\tau}_y h_{\mathrm{BdG}}^*(\mathbf{k})\hat{\tau}_y = -\hat{h}_{\mathrm{BdG}}(-\mathbf{k}),$$ (4.36)

and thus, for a system with particle-hole symmetry, we have

$$\hat{C}h(\mathbf{k})\hat{C}^{-1} = -h(-\mathbf{k}),$$ (4.37)

which should be compared to (4.34). Instead of commuting with the Hamiltonian, if the system is particle-hole symmetric $\hat{C}$ anti-commutes with the Hamiltonian. In terms of the spectrum we have that it is symmetric around the origin: for a particle state with quasi-momentum $\mathbf{k}$ and positive energy E_+ there is a corresponding hole state with quasi-momentum $-\mathbf{k}$ and negative energy $E_- = -E_+$. As for the time-reversal symmetry operator we must have $\hat{C}^2 = \pm 1$ (in the example above we have $\hat{C}^2 = -1$) and $\hat{C}$ is also anti-unitary.

4.1.5.3 Chiral Symmetry

The last discrete symmetry we discuss is the chiral one, also called *sublattice symmetry* as it occurs in systems where the full lattice has a sublattice structure. The chiral symmetry is defined by a unitary operator that anti-commutes with the Hamiltonian. If the system has both time-reversal and particle-hole symmetry we may note that the unitary $\hat{S} = \hat{T}\hat{C}$ anti-commutes with the Hamiltonian, i.e., $\hat{S}\hat{H}\hat{S}^{-1} = -\hat{H}$. Note that $\hat{S}^2 = 1$. A chiral system has a spectrum that is symmetric around $E = 0$, and more precisely for a translationally invariant system the Bloch Hamiltonian transforms as

$$\hat{S}h(\mathbf{k})\hat{S}^{-1} = -h(\mathbf{k}). \tag{4.38}$$

Thus, for a given quasi-momentum $\mathbf{k}$ the states can be paired; one with energy $E_+(\mathbf{k}) > 0$ and one with $E_-(\mathbf{k}) = -E_+(\mathbf{k})$.

Provided that the time-reversal and particle-hole symmetries hold we saw that the system has also a chiral symmetry. The reverse need not be true, i.e., given that the system is chiral, then it need not be time-reversal and particle-hole symmetric. Furthermore, if only one of the time-reversal or particle-hole symmetries hold, we have that the system cannot be chiral symmetric. All in all, you end up with ten possibilities when taking in mind that $\hat{T}$ and $\hat{C}$ can square to $\pm\hat{1}$ and $\hat{S}$ only to $\hat{1}$. This is now known as the ten-fold classification scheme of single particle Hamiltonians [4], which we summarise in Table 4.1.

4.1.6 Topological Invariant

4.1.6.1 Geometric Phase Revisited

Let us recall Sect. 2.3.3. If an initial instantaneous eigenstate $|\Phi(0)\rangle = |\psi_n(\mathbf{R}(0))\rangle$ of some parameter-dependent Hamiltonian $\hat{H} = \hat{H}(\mathbf{R}(t))$ evolves adiabatically around a loop $\mathscr{C} : [0, T] \ni t \mapsto \mathbf{R}(t) | \mathbf{R}(T) = \mathbf{R}(0)$. Then, the state at the final

Table 4.1 The ten different symmetry classes defined with respect to the system's symmetries, which can be time-reversal symmetry (TRS), particle-hole symmetry (PHS), or chiral/sublattice symmetry (CS)

	Cartan label	TRS	PHS	CS
Standard classes	A (unitary)	0	0	0
(Wigner–Dyson)	AI (orthogonal)	+1	0	0
	AII (symplectic)	−1	0	0
Chiral classes	AIII (chiral unitary)	0	0	1
(sublattice)	BDI (chiral orthogonal)	+1	+1	1
	CII (chiral symplectic)	−1	−1	1
BdG classes	D	0	+1	0
	C	0	−1	0
	DIII	−1	+1	1
	CI	+1	−1	1

The notations dates back to a work by Cartan already in 1926

time T is given by

$$|\Phi(T)\rangle = e^{i\gamma_n(\mathcal{C})} \exp\left(-i \int_0^T \varepsilon_n(\mathbf{R}(t))dt\right) |\psi_n(\mathbf{R}(0))\rangle, \qquad (4.39)$$

where $\varepsilon_n(\mathbf{R}(t))$ is the adiabatic energy at $\mathbf{R}(t)$, and the geometric phase can be expressed as the contour integral

$$\gamma_n(\mathcal{C}) = \int_{\mathcal{C}} \mathbf{A}_n(\mathbf{R}) \cdot d\mathbf{R} \qquad (4.40)$$

over the synthetic gauge connection

$$\mathbf{A}_n(\mathbf{R}) = i \langle \psi_n(\mathbf{R})| \nabla_{\mathbf{R}} \psi_n(\mathbf{R})\rangle. \qquad (4.41)$$

For a translationally invariant system we know from Sect. 4.1.1 that the eigenstates have the form (4.8), where the quasi-momentum $\mathbf{k}$ is restricted to the first Brillouin zone. If a constant force $\mathbf{F}$, e.g., caused by an electric field, is applied, and we assume that the energy bands are well separated, it follows from a type of adiabatic approximation (often referred to as the *acceleration theorem*) that $\mathbf{k} \rightarrow \mathbf{k} + \mathbf{F}t$, i.e., the quasi-momentum changes linearly in time. Similarly, if instead a constant magnetic field acts on the systems, then the electrons would perform cyclotron orbits in the Brillouin zone. Such periodic adiabatic motion should generate a geometric phase akin to the Berry phase discussed in Chap. 2. Due to the torus geometry of the Brillouin zone, it is possible to perform a closed loop even in the case where a constant electric field is applied. This is possible already in one dimensions; when the quasi-momentum reaches the Brillouin edge it reappears at the opposite Brillouin edge. The resulting geometric phase is usually referred to as the *Zak phase* [5].

To see how the quasi-momentum **k**, which is a quantum number, may replace the parameter **R** in the original definition of the geometric phase we make the following transformation of the Bloch functions:

$$u_{\mathbf{k}\nu}(\mathbf{x}) = e^{i\mathbf{k}\cdot\mathbf{x}}\psi_{\mathbf{k}\nu}(\mathbf{x}). \tag{4.42}$$

Remember that $u_{\mathbf{k}\nu}(\mathbf{x}) = u_{\mathbf{k}\nu}(\mathbf{x} + \mathbf{R})$, i.e., it is a periodic function with the same periodicity as that of the lattice potential. This function obeys the Schrödinger equation with the transformed Hamiltonian

$$h(\mathbf{k}) = e^{i\mathbf{k}\cdot\mathbf{r}}\hat{H}e^{-i\mathbf{k}\cdot\mathbf{r}} = -\frac{\hbar^2}{2m}(\nabla - i\mathbf{k})^2 + V(\mathbf{x}). \tag{4.43}$$

The quasi-momentum now appears as a parameter of the transformed Hamiltonian, and we may proceed to derive the corresponding geometric phase like in Chap. 2 with the replacements $\mathbf{R} \to \mathbf{k}$ and $|\psi_n(\mathbf{R})\rangle \to |u_\nu(\mathbf{k})\rangle$ (where we have rewritten $u_{\mathbf{k}\nu}(\mathbf{x}) = \langle\mathbf{x}|u_{\mathbf{k}\nu}\rangle = \langle\mathbf{x}|u_\nu(\mathbf{k})\rangle$). Thus, we derive the synthetic gauge connection and curvature

$$\mathbf{A}^{(\nu)}(\mathbf{k}) = i\langle u_\nu(\mathbf{k})|\nabla_{\mathbf{k}}u_\nu(\mathbf{k})\rangle,$$

$$\mathbf{\Omega}^{(\nu)}(\mathbf{k}) = \nabla_{\mathbf{k}} \times \mathbf{A}^{(\nu)}(\mathbf{k}) = i\langle\nabla_{\mathbf{k}}u_\nu(\mathbf{k})| \times |\nabla_{\mathbf{k}}u_\nu(\mathbf{k})\rangle. \tag{4.44}$$

4.1.6.2 Chern and Winding Numbers

In two dimensions, the *Chern number* n_ν for band ν is defined as the Berry phase (divided by 2π) as obtained when integrated over the whole Brillouin zone,

$$n_\nu = -\frac{1}{2\pi}\int_{BZ}\hat{\mathbf{e}}_z \cdot \mathbf{\Omega}^{(\nu)}(\mathbf{k})\,d^2k = -\frac{1}{2\pi}\oint_{\partial BZ}\mathbf{A}^{(\nu)}(\mathbf{k})\cdot d\mathbf{k}, \tag{4.45}$$

where $\hat{\mathbf{e}}_z$ is the unit vector perpendicular to the (k_x, k_y)-plane and ∂BZ is the boundary of the Brillouin zone. Thouless, Kohmoto, Nightingale, and den Nijs showed that for a system with broken time-reversal symmetry, the Hall conductivity σ_{xy} is proportional to the Chern number as

$$\sigma_{xy} = \frac{e^2}{h}\sum_{\nu<\nu_F} n_\nu, \tag{4.46}$$

where e is the electron charge, h Planck's constant, and the sum is over all filled bands [6]. This result is so important that the topological invariant n_ν is sometimes referred to as the TKNN-invariant. As we show next, n_ν is an integer and as such the TKNN result demonstrates that the Hall conductance is quantised in integer multiples of the e^2/h unit.

To show that n_ν must be an integer we first recall that the synthetic gauge structure describes the overall phase factors

$$|u_\nu(\mathbf{k})\rangle \rightarrow e^{if(\mathbf{k})}|u_\nu(\mathbf{k})\rangle \tag{4.47}$$

which transforms the gauge dependent synthetic connection as

$$\mathbf{A}^{(\nu)}(\mathbf{k}) = i\langle u_\nu(\mathbf{k})|\nabla_{\mathbf{k}}u_\nu(\mathbf{k})\rangle + \nabla_{\mathbf{k}}f(\mathbf{k}). \tag{4.48}$$

Assuming that the function $f(\mathbf{k})$ is smooth over the Brillouin zone, the above gauge change cannot affect the Hall conductivity, which is clearly a physical observable. Invoking Stokes' theorem implies that we can find the Hall conductivity by integrating along the boundary of the Brillouin zone, but in a strict mathematical sense the Brillouin zone lacks a boundary since it has the topology of a torus. Hence, we should conclude that the Hall conductivity must vanish! The way out from this paradox is that $\mathbf{A}^{(\nu)}(\mathbf{k})$ need not be well defined throughout the whole Brillouin zone. In fact, if the Chern number is non-zero we must have some singularity in $\mathbf{A}^{(\nu)}(\mathbf{k})$.

The Bloch state vector $|u_\nu(\mathbf{k})\rangle$ has at least two components since we must have at least two bands in a non-trivial situation. If we can find a gauge (4.47) such that $|u_\nu(\mathbf{k})\rangle$ is everywhere smooth we must have, according to the argument above, a vanishing Chern number. One option is to choose a phase factor such that the first component $|u_\nu(\mathbf{k})\rangle_1$ of $|u_\nu(\mathbf{k})\rangle$ becomes real, i.e., $e^{if(\mathbf{k})} = ||u_\nu(\mathbf{k})\rangle_1|/|u_\nu(\mathbf{k})\rangle_1$. If this is possible, then $n_\nu = 0$, so for $n_\nu \neq 0$ the amplitude for the first component $|u_\nu(\mathbf{k})\rangle_1$ must vanish at some points $\mathbf{k}_s$. Naturally, for a zero amplitude, the phase is undefined. At these points, or in some small neighbourhood R_s around them, we therefore need to choose some other gauge. One such choice is to make the second component of the Bloch vector real, $e^{ig(\mathbf{k})} = ||u_\nu(\mathbf{k})\rangle_2|/|u_\nu(\mathbf{k})\rangle_2$. The two gauge choices give two synthetic gauge connections $\mathbf{A}_1^{(\nu)}(\mathbf{k})$ and $\mathbf{A}_2^{(\nu)}(\mathbf{k})$, respectively, and at the boundaries ∂R_s the two are connected via the gauge transformation (cf. (2.64))

$$\mathbf{A}_1^{(\nu)}(\mathbf{k}) = \mathbf{A}_2^{(\nu)}(\mathbf{k}) + \nabla_{\mathbf{k}}\chi(\mathbf{k}), \tag{4.49}$$

where $\chi(\mathbf{k}) = g(\mathbf{k}) - f(\mathbf{k})$. The Chern number can now be calculated as

$$n_\nu = \frac{1}{2\pi}\left(\int_{BZ-R_s}\nabla\times\mathbf{A}_1^{(\nu)}(\mathbf{k})d^2k + \int_{R_s}\nabla\times\mathbf{A}_2^{(\nu)}(\mathbf{k})d^2k\right)$$

$$= \frac{1}{2\pi}\left(\int_{\partial BZ-R_s}\mathbf{A}_1^{(\nu)}(\mathbf{k})\cdot d\mathbf{k} + \int_{\partial R_s}\mathbf{A}_2^{(\nu)}(\mathbf{k})\cdot d\mathbf{k}\right). \tag{4.50}$$

Now, using the fact that the Brillouin zone does not have any boundary and that the two integrals are integrated along the same contour, but in opposite directions we find

$$n_\nu = \frac{1}{2\pi} \int_{\partial R_s} \left(\mathbf{A}_2^{(\nu)}(\mathbf{k}) - \mathbf{A}_1^{(\nu)}(\mathbf{k}) \right) \cdot d\mathbf{k} = \frac{1}{2\pi} \int_{\partial R_s} \nabla \cdot \chi(\mathbf{k}) \cdot d\mathbf{k} = m. \quad (4.51)$$

The last equality follows from the single-valuedness of the wave function. The number m is the *winding number* for the given band. In the example we limited the discussion to a single point $\mathbf{k}_s$, but in general we may have several points and we get a contribution to the total winding number from each of them. It is intuitive to think of the points $\mathbf{k}_s$ as vortices; at the core the wave function vanishes and upon integrating the phase of the wave function around a vortex we must find a multiple of 2π. The integral values of the Chern number reflect its topological nature.

Let us look at a simple non-trivial model example. We consider the two-band Bloch Hamiltonian

$$h(\mathbf{k}) = \mathbf{d}(\mathbf{k}) \cdot \sigma, \quad (4.52)$$

with the vector $\mathbf{d}(\mathbf{k}) = (\sin k_x, \sin k_y, m + \cos k_x + \cos k_y)$ and $\sigma = (\sigma_x, \sigma_y, \sigma_z)$ being the standard Pauli matrices. The eigenenergies are

$$E_\pm(\mathbf{k}) = \pm|\mathbf{d}(\mathbf{k})| = \pm\sqrt{\sin^2 k_x + \sin^2 k_y + (m + \cos k_x + \cos k_y)^2}. \quad (4.53)$$

The two bands become degenerate only for $m = 0, \pm 2$; (1) for $m = 0$ the degeneracy points are $(k_x, k_y) = (0, \pm\pi)$ and $(k_x, k_y) = (\pm\pi, 0)$, (2) for $m = -2$ the degeneracy point is $(k_x, k_y) = (0, 0)$, and (3) for $m = +2$ the degeneracy points are $(k_x, k_y) = (\pm\pi, \pm\pi)$. Three examples of the bands are shown in Fig. 4.3. We

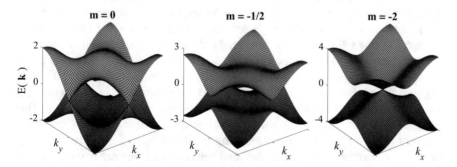

Fig. 4.3 The two bands for the model (4.52), and for three different values of m. There are three different phases according to (4.54), with the transitions as $m = 0$ and $m = \pm 2$. These are topological transitions, where the Chern number changes from the topologically trivial $n_- = 0$ ($|m| > 2$) to the non-trivial $n_- = \pm 1$. The Fermi energy is assumed $E_F = 0$ such that the lowest band is completely filled. At the transition points $m = 0, \pm 2$ the energy gap closes to form CIs or Dirac cones. The gap closing is a characteristic of a topological phase transition

see that the degeneracy points are the counterparts of CIs but in the Brillouin zone, i.e., *Dirac CI's* or *Dirac cones*. For $m = 0$ the CIs are at the Brillouin zone edges and, from the periodicity of the Brillouin zone, we have in total two of them, while for $m = -2$ we have one CI at the centre of the Brillouin zone, and for $m = +2$ (not shown in the figure) the CIs are at the corners and thereby correspond to a single CI. In this example the gauge curvature for the lower band $\Omega^{(-)}(\mathbf{k}) = \frac{1}{2}\frac{\mathbf{d}(\mathbf{k})}{|\mathbf{d}(\mathbf{k})|^3}$ [7]. The Chern numbers as a function of m can be calculated using (4.45) and one finds

$$
n_- = \begin{cases} -1, & -2 < m < 0, \\ +1, & 0 < m < 2, \\ 0, & |m| > 2. \end{cases} \tag{4.54}
$$

For the upper band the Chern number is the same but with the signs swapped ($\sum_\nu n_\nu = 0$ when summed over all bands).

The above result should be contrasted with what we found in the previous chapters—encircling a CI results in a $\pm\pi$ geometric phase irrespective of the distance from the CI, and moreover, if the CI is split, the geometric phase depends on the enclosed area. The synthetic magnetic flux for a CI was restricted to the point of the CI, while when the CI split the flux is non-zero outside the CI point. In the present setting the Chern number is limited to integer values, but at the same time we also define it from integrating over the whole Brillouin zone. If $m \neq 0, \pm2$ there are no Dirac CIs and without changing the Chern numbers, we can define a model $h_{\mathrm{FB}}(\mathbf{k}) = \mathbf{R}(\mathbf{k}) \cdot \boldsymbol{\sigma}$, where $\mathbf{R}(\mathbf{k}) = \mathbf{d}(\mathbf{k})/|\mathbf{d}(\mathbf{k})|$. This is called *band flattening* since the two bands are now simply at $E = \pm1$. Within this formalism it is clear that $\mathbf{R}(\mathbf{k})$ can be viewed as a Bloch vector that lives on the Bloch sphere. Thus, the Brillouin zone has been mapped to a unit sphere. The resulting synthetic magnetic flux is identified as that from a magnetic monopole centred at the origin of the sphere and with charge $\frac{1}{2}$.

We briefly mentioned that the Chern number is a topological invariant. As such it follows that something 'drastic' has to be imposed in order to change it, i.e., we cannot smoothly (adiabatically) deform a band such that its Chern number changes. In order for the Chern number to change we must somehow close the gap between at least two bands. This observation is used to motivate the so called *bulk-boundary correspondence* that link together properties of the bulk with those living on the edges of the material. Assume we have an insulator, i.e., a gap between the conductance and valence band, with a non-vanishing Chern number (or Hall conductance). The surrounding vacuum is a trivial 'insulator', and thus the Chern numbers of the system and its surrounding are different, and the only way to go from one to the other is by closing the band gap. This must happen at the boundary, and we thereby have conducting modes living on the edges. The bulk-boundary correspondence tells that the number of such modes is determined by the Chern number. Going back to Fig. 4.3, we see that the gap closes for $m = 0$ and $m = -2$ (another gap closing occurs for $m = +2$), which are exactly the values that separate the different topologically trivial and non-trivial phases.

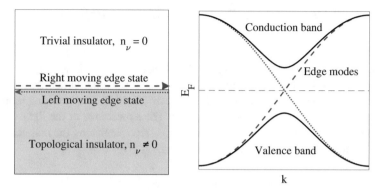

Fig. 4.4 A schematic picture demonstrating the bulk-boundary correspondence. The left plot shows the boundary between an insulator with a non-zero Chern/$\mathbb{Z}_2$ number (topological) and a trivial insulator with a vanishing topological number. At the edge of the topological insulator live conducting states that are topologically protected, i.e., even if the boundary would be deformed or there would be some impurities, these states would still survive. The right plot shows the bulk-boundary spectrum. In the bulk there is a filled valence band separated from an empty conduction band. However, at the boundary the edge states have gapless dispersions shown as a dashed blue and a dotted red line. In a Chern insulator with broken time-reversal symmetry, one finds a chiral edge state moving in either direction, while in a topological insulator with preserved time-reversal symmetry one has counter propagating helical edge states

If time-reversal symmetry is preserved, in two dimensions one has that the Chern number $n_\nu = 0$ as a consequence of the property $\Omega^\nu(-\mathbf{k}) = -\Omega^\nu(\mathbf{k})$ [7]. Hence, it was long believed that breaking of time-reversal symmetry was a necessity for a non-vanishing Hall conductance. However, in 2005 Kane and Mele [8, 9] showed that the picture is in fact more delicate than this. There is a $\mathbb{Z}_2$ topological invariant (with the same notation, the Chern number is a $\mathbb{Z}$ topological number) that can be non-zero for time-reversal symmetric systems in two dimensions. A $\mathbb{Z}_2$ topological insulator has helical edge states instead of the chiral ones when time-reversal symmetry is broken, as depicted in Fig. 4.4. Among these topological insulators we find those called *quantum spin Hall systems*. Here, the average charge current along the edges is zero, but there is instead a spin current running along the edges.

4.2   Spin–Orbit Couplings

Following the previous section on general properties of band theory we continue this chapter by focusing on specific examples where the Dirac CIs appear. In fact, in this section, we do not include any lattice but consider free electrons. We argue how similar coupling terms like those found for JT models in molecules may arise in condensed matter systems. We also demonstrate how such intrinsic spin–orbit couplings give rise to the spin Hall effect.

4.2.1   Rashba and Dresselhaus Spin–Orbit Couplings

The charge carriers in metals are electrons, i.e., spin-1/2 particles. In a magnetic field **B**, the Zeeman energy splitting for the electron is $\mu_B \boldsymbol{\sigma} \cdot \mathbf{B}$ with μ_B the *Bohr magneton*. The spin–orbit coupling (SOC) is a similar splitting, but derives from the electron moving with a momentum **p** in an electric field **E**. The resulting magnetic field due to this motion is $\mathbf{B}_{\text{eff}} \sim \mathbf{E} \times \mathbf{p}/mc^2$. The appearance of the light velocity c in this expression shows that it is a relativistic effect. Replacing the magnetic field in the Zeeman splitting term with the effective field gives us a term[1]

$$\hat{H}_{\text{SO}}(\mathbf{p}) \sim \mu_B (\mathbf{E} \times \hat{\mathbf{p}}) \cdot \boldsymbol{\sigma} / \left(mc^2\right). \tag{4.55}$$

The electric field determines the explicit structure of the spin–orbit coupling, and we differ between *intrinsic* and *extrinsic* SOCs. The intrinsic SOC derives from the electric field due to the atoms forming the lattice. This is normally a weak perturbation to the potential term. The extrinsic SOC appears due to any external field, either a manually applied field or if the sample is attached to some other material. Since the SOC is linear in the momentum, we have $\hat{H}_{\text{SO}}(-\mathbf{p}) = -\hat{H}_{\text{SO}}(\mathbf{p})$. This does not, however, imply breaking of time-reversal symmetry since we have half-integer spins, yielding

$$\hat{T}\hat{H}_{\text{SO}}(\mathbf{p})\hat{T}^{-1} = \hat{T}\alpha_R(\mathbf{e}_z \times \hat{\mathbf{p}}) \cdot \boldsymbol{\sigma}\,\hat{T}^{-1} = \alpha_R(\mathbf{e}_z \times (-\hat{\mathbf{p}})) \cdot (-\boldsymbol{\sigma}) = \hat{H}_{\text{SO}}(\mathbf{p}), \tag{4.56}$$

where we introduced the *Rashba parameter* α_R and took the electric field to be in the z-direction. With this electric field we have the Rashba SOC

$$\hat{H}_{\text{RSO}}(\mathbf{p}) = \alpha_R(\hat{p}_y\sigma_x - \hat{p}_x\sigma_y). \tag{4.57}$$

There is another type of SOC called Dresselhaus and is given by

$$\hat{H}_{\text{DSO}}(\mathbf{p}) = \beta_D(\hat{p}_x\sigma_x - \hat{p}_y\sigma_y). \tag{4.58}$$

The symmetries of the material determine which kind of SOC that is supported. The strengths α_R and β_D are typically of the same order. Theoretically the two types of SOC are equivalent in the sense that there exists a unitary transformation mapping one into the other (π rotation around $\sigma_x + \sigma_y$).

---

[1] The derivation given here for the Rashba term is non-rigorous, but it actually results in the correct expression.

Due to the unitary equivalence between the two, here we consider only the Rashba SOC. The Hamiltonian in the absence of any external potential is simply

$$\hat{H}_{SOC}(\mathbf{p}) = \frac{\hat{p}_x^2}{2m} + \frac{\hat{p}_y^2}{2m} + \alpha_R(\hat{p}_y\sigma_x - \hat{p}_x\sigma_y). \tag{4.59}$$

Let (p_x, p_y) be eigenvalues of $(\hat{p}_x, \hat{p}_y)$. In polar coordinates $p = \sqrt{p_x^2 + p_y^2}$ and $\phi = \arctan(p_x/p_y)$), we have the eigenvalues

$$E_\pm(p) = \frac{p^2}{2m} \pm \alpha_R p \tag{4.60}$$

and corresponding eigenvectors

$$|\psi_\pm(\mathbf{p})\rangle = \frac{1}{\sqrt{2}} \begin{bmatrix} 1 \\ \mp i e^{i\phi} \end{bmatrix}. \tag{4.61}$$

The Bloch vectors for the eigenstates are $\mathbf{R}_\pm(\mathbf{p}) = (\langle\sigma_x\rangle, \langle\sigma_y\rangle, \langle\sigma_z\rangle) = (\pm\sin\phi, \mp\cos\phi, 0)$, which show how the spin orients for a given momentum. In Fig. 4.5a, we show the typical eigenenergies of the Rashba Hamiltonian. At the origin $(p_x, p_y) = (0, 0)$ an SOC CI appears and like for the linear version $(g = 0)$ of the JT models described in Sect. 3.3.2, the lower energy branch has a degenerate global energy minimum for $p = 2\alpha_R$. In Fig. 4.5b we show instead how

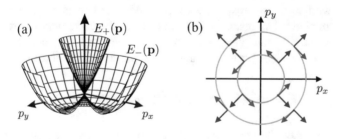

Fig. 4.5 The dispersions (eigenenergies) for the Rashba Hamiltonian (4.59) shown in (**a**). The lower branch $E_-(\mathbf{p})$ displays the characteristic sombrero shape, while the upper $E_+(\mathbf{p})$ possesses a global minimum at the Dirac cone. These energies should be compared with the adiabatic potential surfaces $V_\pm^{(ad)}(\mathbf{x})$ of the linear version $(g = 0)$ of the $E \times \varepsilon$ JT model described in Sect. 3.3.2. Naturally, these surfaces are the true eigenenergies of the Hamiltonian, contrary to the adiabatic potential surfaces discussed extensively in Chap. 3. Nonetheless, the presence of the CI in both is evident. The right plot (**b**) demonstrates the locking of the spin with the momentum; the red arrows give the directions of the spin and the blue ones the directions of the momentum. The two grey circles represent a constant energy cut through the two branches $E_\pm(\mathbf{p})$. The important observation in (**b**) is the perpendicular locking between spin and momentum, and that the right angle is opposite between the two branches. Another manifestation of this chiral effect can be traced to the synthetic magnetic flux that point in either positive or negative p_z direction depending on the spin, see (4.64)

the spin (red arrows) locks to the momentum. As is clear from the Bloch vector, for the upper/lower branch we see an anti-clockwise or a clockwise orientation, respectively.

4.2.2 Intrinsic Spin Hall Effect

The SOC term gives rise to many phenomena [10] of which the spin Hall effect [11] is without doubt one of the most important ones. The Hamiltonian (4.59) may be rewritten as

$$
\hat{H}_{SOC}(\mathbf{p}) = \frac{(\hat{p}_x - \hat{A}_x)^2}{2m} + \frac{(\hat{p}_y - \hat{A}_y)^2}{2m} + \hat{\Phi},
\tag{4.62}
$$

where $\hat{A}_x = m\alpha_R\hat{\sigma}_y$, $\hat{A}_y = -m\alpha_R\hat{\sigma}_x$, and $\hat{\Phi} = -\alpha_R^2 m\hat{1}$, with $\hat{\sigma}_x, \hat{\sigma}_y$ Pauli operators. Following the discussion of Sect. 2.3.1 we can interpret the SOC as a synthetic vector potential $\hat{\mathbf{A}} = (\hat{A}_x, \hat{A}_y)$. Recall (2.35), i.e.,

$$
\hat{F}_{kl} = \partial_k \hat{A}_l - \partial_l \hat{A}_k - i[\hat{A}_k, \hat{A}_l], \qquad k, l = x, y
\tag{4.63}
$$

for the curvature tensor. The first two terms clearly vanish, while the third one, stemming from the non-Abelian structure of the vector potential, is non-zero. With the given vector potential we find that the corresponding operator-valued synthetic magnetic field becomes $\hat{\mathbf{B}} = (0, 0, m^2\alpha_R^2\hat{\sigma}_z)$ such that the electrons with spin up and those with spin down experience a perpendicular constant magnetic field in opposite directions. The resulting Lorentz force is therefore opposite for the two spin components. We may also derive the force from the Heisenberg equations of motion, $\partial_t \hat{O} = -i[\hat{O}, \hat{H}]$, which results in the spin-dependent force (cf. Sect. 3.4)

$$
\hat{\mathbf{F}} = m\frac{d^2}{dt^2}\hat{\mathbf{r}} = m[\hat{H}, [\hat{\mathbf{r}}, \hat{H}]] = m\alpha_R^2(\hat{\mathbf{p}} \times \mathbf{e}_z)\hat{\sigma}_z.
\tag{4.64}
$$

An electron in, say, the spin state $|\uparrow\rangle$ and with an initial velocity along the x-direction experiences a force in the negative y-direction, and vice versa for an electron in the state $|\downarrow\rangle$. Thus, the Rashba SOC manifests in a spin-dependent splitting of the electronic motion, similar to the splitting in a Stern–Gerlach setting. This observation explains why SOC systems are of interest for *spintronics*, where the spin degree of freedom of the electron is utilised for various practical purposes like data storage.

For the given Rashba Hamiltonian (4.59), the motion of the two spin components of the electron will not simply display simple cyclotron motion (due to the coupling of the motion to the spin). A typical example of the evolution is shown in Fig. 4.6, which gives the expectation value of the position for the two spin components. The initial wave packet is a Gaussian with a non-zero velocity in the x-direction and

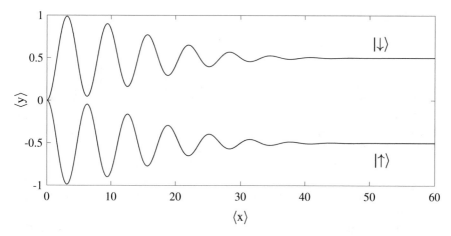

Fig. 4.6 Demonstration of the intrinsic spin Hall effect. We consider an initial Gaussian state $\psi(p_x, p_y, t = 0) = \phi_\downarrow(p_x, p_y, 0)|\downarrow\rangle + \phi_\uparrow(p_x, p_y, 0)|\uparrow\rangle$ with $\phi_\downarrow(p_x, p_y, 0) = \phi_\uparrow(p_x, p_y, 0) = \sqrt{1/2\pi\sigma^2} \exp\left[-(p_x - 1)^2/2\sigma^2 - p_y^2/2\sigma^2\right]$. In our scaled units ($\hbar = m = 1$) we have taken the width $\sigma = 0.1$ and the initial momentum $(p_x, p_y)_0 = (1, 0)$ such that initially there is no motion along the y-direction. The figure shows the evolution of the average positions, $\langle x \rangle$ and $\langle y \rangle$, of the respective spin component. The early oscillations is a manifestation of so called *Zitterbewegung*, and their relaxation results from the dephasing among the different momentum components of the initial Gaussian state (for an initial plane wave the oscillations would be indefinite). The asymptotic value of the $\langle y \rangle$ expectations is $\pm\frac{1}{2}p_{x0}$ (depending on the spin state), i.e., the slower the initial state the more displaced wave packets in the transverse direction. It is this spin-dependent transverse motion that gives a net transverse spin-current—the intrinsic spin Hall effect

an equal weight of the two spin components (see caption for details). The initial splitting of the wave packet is followed by an oscillatory phase, which however collapses and $\langle y \rangle$ tends to an asymptotic value. With the two spin components equally populated we have no net transverse charge current, while instead there is a net transverse spin current. In a finite sample we expect a predominance of up-spins on one edge and down-spins on the opposite edge. This is the intrinsic spin Hall effect, and it is 'intrinsic' since it derives from the SOC that is a characteristic of the material, and not from some externally applied field (like in the normal Hall effect). Note also that time-reversal symmetry is preserved in this system.

An external magnetic field would typically imply a term $\sim \hat{\sigma}_z$ in the Hamiltonian. This would break the time-reversal symmetry, and there would no longer be an equal balance between the populations of the two spin components as the system evolves. This means that, in addition to the transverse spin current, there would also be a net transverse charge current. This is the setting of the *intrinsic anomalous Hall effect*.

4.3 Superconductors

This section continues with a system where Dirac CIs emerge, namely in p-wave superconductors. A crucial difference to other systems considered in this chapter is that the CI is induced by interactions, and appears at the mean-field level.

As the temperature is decreased, some metals become superconducting below a critical temperature $T_c > 0$. Thus, they become perfect conductors so that the resistivity vanishes. Being fermions, the electrons cannot condense unless pairs of electrons bound into bosons, called *Cooper pairs*. It turns out that the phonons arising from vibrations in the lattice of ions can mediate the effective attractive interaction needed for Cooper pair formation. Any interaction will scatter two electrons with certain momenta and spins into two electrons with new momenta and spins. However, at low temperatures the states deep below the Fermi surface are occupied so that scattering into those states is forbidden. Likewise, we expect that states with energies much larger than the Fermi energy are unoccupied. In other words, we may restrict the scattering processes to those involving states in the close vicinity of the Fermi surface. Moreover, if we consider scattering between electrons with opposite momenta, $\mathbf{k}$ and $-\mathbf{k}$, it follows that the centre of mass momentum of the two electrons is zero so that the Cooper pair carries no kinetic energy. Finally, due to the anti-symmetry of the electron wave function the scattered electrons must also have opposite spins. Thus, the relevant two-electron process is

$$|\mathbf{k}\uparrow, -\mathbf{k}\downarrow\rangle \rightarrow |\mathbf{k}'\uparrow, -\mathbf{k}'\downarrow\rangle, \tag{4.65}$$

where it is understood that the ket represents the state of the two electrons in the momentum representation, and the momenta $\mathbf{k}$ and $\mathbf{k}'$ are restricted to a thin shell around the Fermi momentum k_F. The corresponding (second quantised) Hamiltonian becomes

$$\hat{H} = \sum_{\mathbf{k},\sigma} \epsilon(\mathbf{k}) \hat{c}_{\mathbf{k}\sigma}^\dagger \hat{c}_{\mathbf{k}\sigma} - \sum_{\mathbf{k},\mathbf{k}'} V_{\mathbf{k}\mathbf{k}'} \hat{c}_{\mathbf{k}'\uparrow}^\dagger \hat{c}_{-\mathbf{k}'\downarrow}^\dagger \hat{c}_{\mathbf{k}\uparrow} \hat{c}_{-\mathbf{k}\downarrow}. \tag{4.66}$$

Here, $\hat{c}_{\mathbf{k}\sigma}^\dagger$ ($\hat{c}_{\mathbf{k}\sigma}$) creates (annihilates) an electron with momentum $\mathbf{k}$ and spin σ. The scattering amplitude $V_{\mathbf{k}\mathbf{k}'}$ is determined from the shape of the interacting potential $V(|\mathbf{r}_i - \mathbf{r}_j|)$ between the electrons at positions $\mathbf{r}_i$ and $\mathbf{r}_j$. At the mean-field level, we decouple the two momentum modes by introducing the *superconducting order parameter*

$$\Delta_{\mathbf{k}} = \sum_{\mathbf{k}'} V_{\mathbf{k}\mathbf{k}'} \langle \psi_0 | \hat{c}_{\mathbf{k}'\uparrow}^\dagger \hat{c}_{-\mathbf{k}'\downarrow}^\dagger | \psi_0 \rangle, \tag{4.67}$$

where $|\psi_0\rangle$ is the self-consistently obtained mean-field ground state. Within the mean-field approach we derive the *Bogoliubov–de Gennes Hamiltonian*

$$\hat{H}_{\text{BdG}} = \sum_{\mathbf{k}} \left[\epsilon(\mathbf{k}) \hat{c}_{\mathbf{k}\sigma}^{\dagger} \hat{c}_{\mathbf{k}\sigma} - \left(\Delta_{\mathbf{k}} \hat{c}_{\mathbf{k}\uparrow} \hat{c}_{-\mathbf{k}\downarrow} + \Delta_{\mathbf{k}}^* \hat{c}_{-\mathbf{k}\downarrow}^{\dagger} \hat{c}_{\mathbf{k}\uparrow}^{\dagger} \right) \right], \tag{4.68}$$

or when expressed in terms of the *Nambu spinors*

$$\hat{\Psi}_{\mathbf{k}}^{\dagger} = \left[\hat{c}_{\mathbf{k}\uparrow}^{\dagger}, \, \hat{c}_{-\mathbf{k}\downarrow} \right], \qquad \hat{\Psi}_{\mathbf{k}} = \begin{bmatrix} \hat{c}_{\mathbf{k}\uparrow} \\ \hat{c}_{-\mathbf{k}\downarrow}^{\dagger} \end{bmatrix}, \tag{4.69}$$

one finds (up to an overall constant term)

$$\hat{H}_{\text{BdG}} = \sum_{\mathbf{k}} \hat{\Psi}_{\mathbf{k}}^{\dagger} \begin{bmatrix} \epsilon(\mathbf{k}) & -\Delta_{\mathbf{k}}^* \\ -\Delta_{\mathbf{k}} & -\epsilon(\mathbf{k}) \end{bmatrix} \hat{\Psi}_{\mathbf{k}} \equiv \sum_{\mathbf{k}} \hat{\Psi}_{\mathbf{k}}^{\dagger} h(\mathbf{k}) \hat{\Psi}_{\mathbf{k}}. \tag{4.70}$$

What about the $\mathbf{k}$-dependence of the order parameter $\Delta_{\mathbf{k}}$? As mentioned above, this depends on the actual interaction potential between the scattered pair of electrons. If s-wave scattering dominates we have no spatial dependence of the order parameter and $\Delta_{\mathbf{k}}$ becomes just a constant. In a 'p-wave superconductor' instead, the p partial wave scattering dominates. We could, for example, imagine the scenarios

$$\Delta_{\mathbf{k}} = \begin{cases} \alpha(k_x + i k_y), \\ \alpha(k_x - k_y). \end{cases} \tag{4.71}$$

The first one preserves chiral symmetry (often referred to as a *chiral superconductor*), but breaks time-reversal symmetry. The corresponding Bloch Hamiltonian with $\epsilon(\mathbf{k}) = k_x^2/2 + k_y^2/2$ is simply

$$h_p(\mathbf{k}) = \left(\frac{k_x^2}{2} + \frac{k_y^2}{2} \right) \sigma_z - \alpha k_x \sigma_x - \alpha k_y \sigma_y, \tag{4.72}$$

with the eigenvalues

$$\varepsilon_{\pm}^{(1)}(\mathbf{k}) = \pm \sqrt{\left(\frac{k_x^2}{2} + \frac{k_y^2}{2} \right)^2 + \alpha^2 (k_x^2 + k_y^2)}. \tag{4.73}$$

For small $\mathbf{k}$ we regain a linear Hamiltonian and the energies form a Dirac CI at $(k_x, k_y) = (0, 0)$. Note the similarity with the Rashba SOC Hamiltonian (4.59).

For the second example of (4.71), when time-reversal symmetry is not broken, the eigenvalues are

$$\varepsilon_\pm^{(2)}(\mathbf{k}) = \pm \sqrt{\left(\frac{k_x^2}{2} + \frac{k_y^2}{2}\right)^2 + \alpha^2 (k_x - k_y)^2}. \tag{4.74}$$

Again we find a degeneracy for $(k_x, k_y) = (0, 0)$, but not a CI. The geometric phase when encircling this degeneracy vanishes.

Another important class of superconductors are the d-wave ones. For example, all the known high-T_c superconductors belong to this class. Here the $\mathbf{k}$-dependence of the order parameter is quadratic, e.g.,

$$\Delta_\mathbf{k} = \begin{cases} \alpha(k_x^2 - k_y^2), \\ \alpha(k_x^2 - k_y^2 + ik_x k_y). \end{cases} \tag{4.75}$$

As a result, the intersections cannot be conical like for p-wave superconductors. The physics of such quadratic intersections, or *glancing intersections*, was discussed in Sect. 3.3.2 in terms of the Renner–Teller model.

4.4   Graphene

A well-known example of Dirac CIs are those appearing in the dispersions of graphene, i.e., single sheets of carbon atoms forming a hexagonal lattice. The band structure has been known since long, and it was early on understood that electronic properties of this material should be very interesting. However, it was believed that such a two-dimensional sheet of carbon atoms would not be stable against fluctuations, and it was also considered extremely hard, if not impossible, to manufacture clean and large enough two-dimensional sheets. The story how the first graphene sheets were extracted using graphite and scotch tape is well-known, and it is a good example of how some imagination can take you far, even as far as a Nobel prize.

4.4.1   Tight-Binding Band Spectrum

Since graphene is such an important system in the context of Dirac CIs, let us start by outlining how the hexagonal structure emerges when carbon atoms form a two-dimensional lattice. The six electrons of carbon configure as $(1s)^2(2s)^2(2p)^2$. The inner two electrons fill the first shell and do not contribute to the chemical bonding between carbon atoms. When the carbon atoms come close and form molecules the

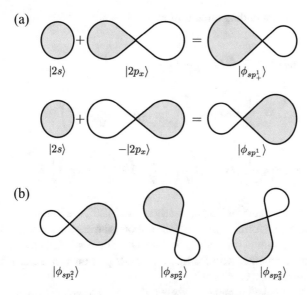

Fig. 4.7 Schematic picture of the sp-hybridisation. In molecules like acetylene (C_2H_2), one $2s$-orbital state hybridises with a $2p$-orbital state to form the states $|\phi_{sp_\pm^1}\rangle$ of (4.76). This is visualised in (**a**). In graphene, the $2s$-orbital hybridises with two $2p$-orbital states according to (4.77). The resulting orbitals are shown in (**b**), and it is seen that they arrange with a mutual 120° rotation. This 120° order causes the hexagonal lattice structure of graphene

electronic energies shift and the $2s$ and $2p$ electrons hybridise. This hybridisation is the reason for the formation of molecules, like hydrocarbons and fullerenes, and it is also the reason for the hexagonal lattice structure of graphene. For several hydrocarbons, one s- and one p-orbital hybridise as

$$|\phi_{sp_\pm^1}\rangle = \frac{1}{\sqrt{2}}\left(|2s\rangle \pm |2p_x\rangle\right), \qquad (4.76)$$

where $|2s\rangle$ and $|2p_\alpha\rangle$ ($\alpha = x, y, z$) are the corresponding quantum states. The two sp^1-orbitals of the hybridised states $|\phi_{sp_\pm^1}\rangle$ are visualised in Fig. 4.7a. In for example acetylene, the two $|\phi_{sp_\pm^1}\rangle$ form a σ-bond; the two orbitals, localised around each carbon atom, greatly overlap to form the bond. The remaining two $2p$-orbitals, p_y and p_z, which are perpendicular to the symmetry axis, constitute two π-bonds that are weaker in strength since the overlap is smaller. Of course, it is also possible for a $2s$-orbital to hybridise with two $2p$-orbitals, which happens for example in benzene. The corresponding three orthonormal quantum states are

$$|\phi_{sp_1^2}\rangle = \frac{1}{\sqrt{3}}|2s\rangle - \sqrt{\frac{2}{3}}|2p_x\rangle,$$

$$|\phi_{sp_2^2}\rangle = \frac{1}{\sqrt{3}}|2s\rangle + \sqrt{\frac{2}{3}}\left(\frac{\sqrt{3}}{2}|2p_y\rangle + \frac{1}{2}|2p_x\rangle\right), \qquad (4.77)$$

$$|\phi_{sp_3^2}\rangle = \frac{1}{\sqrt{3}}|2s\rangle - \sqrt{\frac{2}{3}}\left(-\frac{\sqrt{3}}{2}|2p_y\rangle + \frac{1}{2}|2p_x\rangle\right).$$

These lie in the xy-plane and are mutually rotated 120°, see Fig. 4.7b. A result of the rotation is that when forming a lattice, the σ-bonds arrange the atoms in a hexagonal shape. The remaining p_z-orbitals give three additional π-bonds for every hexagon. The six σ-bonds together with the three π-bonds give an average distance between the carbon atoms to be 0.142 nm.

Graphite is stacked graphene sheets that hold together via van der Waals interaction. The weak van der Waals interaction causes graphite to be fragile, which is also why it is used in pencils. By combining carbon hexagons with pentagons, the graphene sheet curves and can give rise to fullerenes; C_{60} being the most known example. Rolled sheets can give carbon nanotubes which typically have a diameter of several nanometers (the diameter of C_{69}, $d \sim 0.7$ nm) to be compared to the 0.142 nm lattice spacing of graphene.

The hexagonal lattice is formed from two triangular sublattices, A and B, as shown in Fig. 4.8a as the blue and red dots. Nearest neighbour bonds are between

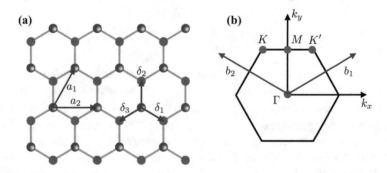

Fig. 4.8 The hexagonal graphene lattice (**a**) with blue and red dots forming the two triangular sublattices A and B. The vectors a_1 and a_2 are Bravais primitive lattice vectors for sublattice A, while the three δ_i vectors denote nearest neighbours. The reciprocal lattice vectors, b_1 and b_2, are shown in (**b**). These define the first Brillouin zone, which for a triangular lattice is a hexagon rotated 30° with respect to the hexagons of the graphene lattice. Also marked in (**b**) are the K and M points, as well as the Γ point. At the K points the two bands touch and form a Dirac CI

the two sublattices, while the next nearest neighbours belong to the same sublattice. The Bravais vectors for one sublattice,

$$a_1 = a\hat{e}_x, \qquad a_2 = \frac{a}{2}\left(\hat{e}_x + \sqrt{3}\hat{e}_y\right), \tag{4.78}$$

with a the lattice spacing, are given in the figure. Note that graphene is an example of a system with two atoms per unit cell. The reciprocal lattice vectors b_1 and b_2, defined by $b_i \cdot a_j = 2\pi \delta_{ij}$, span the reciprocal lattice and determine the first Brillouin zone which we display in Fig. 4.8b. As we see, the reciprocal lattice of a triangular lattice is a hexagonal lattice.

Let us for now on consider spinless (polarised) fermions. Given that there are two atoms per unit cell, the tight-binding Hamiltonian is

$$\hat{H} = -t \sum_{\langle ij \rangle} \left(\hat{a}_i^\dagger \hat{b}_j + h.c.\right), \tag{4.79}$$

where t is the tunnelling amplitude between nearest neighbours (i.e., between the two sublattices A and B) such that the $\hat{a}_i^\dagger$ creates an electron at site i on sublattice A and similarly for $\hat{b}_j^\dagger$ on sublattice B. To diagonalise the Hamiltonian we follow the procedure of Sect. 4.1.4, i.e., we go to the momentum representation and find the corresponding Bloch Hamiltonian. Let $\mathbf{R}_i$ denote the position of the ith site in sublattice A, and δ_i the vector connecting neighbouring sites in the A and B sublattices (see Fig. 4.8a). We express the momentum representation of the electronic operators in the following form:

$$\begin{bmatrix} \hat{a}_{\mathbf{k}} \\ \hat{b}_{\mathbf{k}} \end{bmatrix} = \sum_i \exp(i\mathbf{k} \cdot \mathbf{R}_i) \begin{bmatrix} \hat{a}_i e^{-i\mathbf{k}\cdot\delta_2/2} \\ \hat{b}_i e^{i\mathbf{k}\cdot\delta_2/2} \end{bmatrix}. \tag{4.80}$$

Here the i-subscript represents the unit cell i which contains the two electron operators. When substituted in the Hamiltonian (4.79) we derive the Bloch Hamiltonian

$$h(\mathbf{k}) = \begin{bmatrix} 0 & \Delta_{\mathbf{k}} \\ \Delta_{\mathbf{k}}^* & 0 \end{bmatrix}, \tag{4.81}$$

where

$$\Delta_{\mathbf{k}} = -t \sum_{i=1}^{3} e^{i\mathbf{k}\cdot\delta_i} = -te^{-ik_x a}\left[1 + 2e^{i3k_x a/2}\cos\frac{\sqrt{3}}{2}k_y a\right]. \tag{4.82}$$

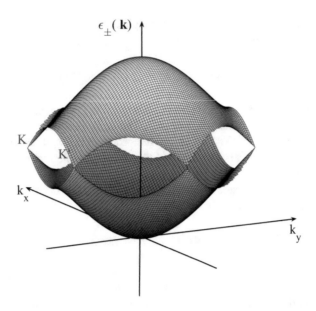

Fig. 4.9 The energy dispersions (4.83) plotted within the first Brillouin zone. The Dirac CI at the six corners are clearly visible. For every unit cell in the reciprocal lattice there are in total two Dirac CI, indicated here at the K-points, K and K'

The block form of the Hamiltonian (4.81) implies a chiral symmetry; $h(\mathbf{k})$ anti-commutes with σ_z. The eigenvalues are

$$\epsilon_\pm(\mathbf{k}) = \pm|\Delta_\mathbf{k}| = \pm\sqrt{1 + 4\cos\left(\frac{3k_x a}{2}\right)\cos\left(\frac{\sqrt{3}k_y a}{2}\right) + 4\cos^2\left(\frac{\sqrt{3}k_y a}{2}\right)}. \tag{4.83}$$

In Fig. 4.9 we plot the two energy dispersions within the first Brillouin zone. At the corners of the Brillouin zone the two dispersions become degenerate and form Dirac CIs. In total there are six such points, but to every unit cell in the reciprocal lattice there are only two Dirac CIs as is evident from the reciprocal lattice vectors b_1 and b_2 of Fig. 4.8b. These two points are called K-points and are marked in the figure. It is arbitrary which two nearby Dirac CIs to choose for the K-points. The ones of the figure are located at $\mathbf{q}_\pm = (\pm\frac{2\pi a}{3\sqrt{3}}, -\frac{2\pi a}{3})$. If we linearise the $\Delta_\mathbf{k}$ around these points we find

$$\Delta_K(\mathbf{q}) = \hbar v_F(q_x + iq_y) = \Delta_{K'}^*(\mathbf{q}), \tag{4.84}$$

where the *Fermi velocity* $v_F = 3ta/2\hbar$, which for graphene is of the order 10^6 m/s. The reason to call it Fermi velocity is because with one electron per site we have half filling (since electrons are spin-1/2 particles), and the Fermi surface is exactly at the Dirac CIs. The degeneracy at these points implies that the density of states vanishes at the Fermi energy, and even though the system is not gapped graphene is what is called a *semimetal*—the conductance and valence bands touch at single points. The Dirac CIs in graphene have been experimentally explored, with focus on how interactions influence the relativistic behaviour [12]. A detailed experimental study of the dispersions of the hexagonal lattice characteristic of graphene has also been performed in terms of ultracold fermionic atoms in optical lattices [13].

4.4.2 Relativity at Almost 'Zero'

At very low temperatures, the physics close to the Fermi energy $\epsilon_F = 0$ determine the behaviour of the system. If we focus on the K point and linearise the dispersion, we can write the effective Bloch Hamiltonian

$$h_{\text{Deff}}(\mathbf{q}) = \hbar v_F \begin{bmatrix} 0 & q_x + iq_y \\ q_x - iq_y & 0 \end{bmatrix}. \tag{4.85}$$

In the linear approximation, we thereby derive a two-dimensional Dirac equation (which is the reason why the CIs appearing in momentum space are called Dirac cones). Note in particular the absence of a 'mass' term in the Dirac equation; such a term would open up a gap at the Dirac CIs. We conclude that the electrons should behave as massless relativistic particles with a 'speed of light' v_F that is about 300 times smaller than c. The other Dirac CI, the K' point, is identical up to complex conjugation of $h_{\text{eff}}(\mathbf{q})$, which demonstrates the different 'helicity' of the two points.

A legitimate question is whether the Dirac CIs are a result of the tight-binding approximation, i.e., would coupling terms beyond nearest neighbours open up a gap in the spectrum? Next nearest neighbour tunnelling terms induce hopping between sites within the same sublattice. If the corresponding tunnelling amplitude is s, we should add to our tight-binding Hamiltonian (4.79) the term

$$\hat{H}_{nnn} = -s \sum_{\langle\langle ij \rangle\rangle} \left(\hat{a}_i^\dagger \hat{a}_j + \hat{b}_i^\dagger \hat{b}_j + h.c. \right), \tag{4.86}$$

where the sum is over next nearest neighbours. Since the tunnelling amplitude is identical for the two sublattices it only results in an overall energy shift for the two bands. Hence, it adds a term

$$f(\mathbf{k}) = -s \left[2\cos\left(\sqrt{3}k_y a\right) + 4\cos\left(\frac{\sqrt{3}k_y a}{2}\right) \cos\left(\frac{3k_x a}{2}\right) \right] \tag{4.87}$$

to the diagonal of the Bloch Hamiltonian (4.81). This will not split the Dirac CIs, but alter the shape of the bands. Adding even longer hopping terms will also not break the Dirac CI degeneracy. In fact, the degeneracy is protected due to time-reversal and inversion symmetry. The time-reversal symmetry manifests as $h^*(\mathbf{k}) = h(-\mathbf{k})$, while inversion symmetry implies invariance of the model under the sublattice exchange $A \leftrightarrow B$. So if we would have different next nearest neighbour tunnelling amplitudes between the two sublattices we would break the inversion symmetry and open up a gap in the spectrum. Similarly, we could shift the onsite energies between the A and B sites, which would also lead to a Dirac CI splitting and a mass term in the Dirac equation.

4.4.3 The Haldane Model

We have seen that for a Chern insulator, the bulk is gapped and the edges host gapless edge states. By breaking the inversion symmetry we would indeed have a gapped bulk, but the model would be topologically trivial. This could be achieved by including a 'mass' term $m\sigma_z$ to the graphene Bloch Hamiltonian (4.81). One may ask whether there is a way to open up a gap while preserving the inversion symmetry, and find a topologically non-trivial model. This is achieved by breaking the time-reversal symmetry as suggested by Haldane [14]. He introduced next nearest neighbour hopping terms with complex tunnelling amplitudes (4.79).

As mentioned in the previous section, next nearest neighbouring tunnelling only induces hopping within one and the same sublattice, as can be seen in (4.86). In the Haldane model the next nearest neighbour tunnelling amplitude $s = it_2$, which we schematically depict in Fig. 4.10. Due to the Hermiticity of the Hamiltonian the tunnelling amplitudes in opposite directions are their own complex conjugates, i.e.,

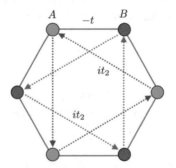

Fig. 4.10 The structure of the tunnellings in the Haldane model: The nearest neighbour hopping between the two sublattices (blue and red dots) occurs with a real tunnelling amplitude $-t$, while the next nearest neighbour hopping has a purely imaginary tunnelling amplitude it_2. In the direction of the arrows the tunnelling amplitude is it_2 such that in the other direction it is $-it_2$

in the figure the tunnelling amplitude is it_2 for hopping in the direction of the arrow, and it is $-it_2$ for hopping in the opposite direction.

The complex Hamiltonian in real space implies that the time-reversal symmetry has been broken. This becomes clear by noting that the additional next nearest neighbouring tunnelling terms would add $2t_2 \sum_i \sigma_z \sin(\mathbf{k} \cdot \mathbf{b}_i)$ to the Bloch Hamiltonian (4.81), where the $\mathbf{b}_i$ vectors are either of the three blue or three red arrows shown in Fig. 4.10. Under time-reversal swapping (cf. (4.34)), this term flips sign and thereby breaks time-reversal symmetry. An alternative way of understanding the lack of time-reversal symmetry is by ascribing a (synthetic) magnetic field to the complex tunnelling terms as described in Appendix A.2. The magnetic flux φ through one of the triangles in Fig. 4.10 is $\varphi = \mathrm{Arg}\left[(-it_2)^3\right] = \pi/2$.

Let us include both the 'mass' term $m\sigma_z$ and the complex Haldane term. With $t_2 = 0$ the two Dirac CIs split and the corresponding Dirac equation (4.85) acquires a mass term. The system is topologically trivial with no edge states. When t_2 is non-zero we have that for small tunnelling amplitudes the gap persists, but when $|t_2| > m/3\sqrt{3}$ one of the two Dirac CI closes. At this point the system is no longer topologically trivial and chiral edge states appear. Features of the topology have been analysed in an optical lattice experiment [15]. In such systems the complex next nearest neighbour hopping has been engineered with the technique of 'lattice shaking'.

4.5 Weyl Semimetals

Recall from above that a semimetal is characterised by the valence and conduction bands touching at a point such that the density of states vanishes for this energy; for graphene this coincides with the Fermi energy. As the spectrum is not gapped, it is not an insulator, and in particular not a topological insulator. Nevertheless, there is a non-trivial synthetic gauge curvature associated with the corresponding Dirac CI— like for the general case of a CI, $\gamma = \pi$. In the vicinity of the Dirac CI the dispersion is approximately linear and the effective low energy physics could be explained in terms of the Dirac equation (4.85). A natural question that comes to mind is if there is something corresponding to graphene in three dimensions with Dirac CIs now in all three momenta $\mathbf{k} = (k_x, k_y, k_z)$. As we will see, this is possible and such materials have been named *Weyl semimetals* [16].

The fact that there are three Pauli matrices implies that Dirac CIs in three dimensions are allowed. A generic two-band Bloch Hamiltonian in three dimensions would take the form

$$h_W(\mathbf{k}) = f_0(\mathbf{k})\mathbb{I} + f_1(\mathbf{k})\sigma_x + f_2(\mathbf{k})\sigma_y + f_3(\mathbf{k})\sigma_z \tag{4.88}$$

for some real functions $f_\alpha(\mathbf{k})$ ($\alpha = 0, 1, 2, 3$). The eigenvalues are $\epsilon_\pm(\mathbf{k}) = f_0(\mathbf{k}) \pm \sqrt{f_1^2(\mathbf{k}) + f_2^2(\mathbf{k}) + f_3^2(\mathbf{k})}$ such that a Dirac CI means that $f_1(\mathbf{k}) = f_2(\mathbf{k}) =$

$f_3(\mathbf{k}) = 0$ simultaneously, which is a $\mathbf{k}$-space analogue of (3.3). These functions are planes in the $\mathbf{k}$-space, and any of them will typically be zero along a line. Furthermore, $f_i(\mathbf{k}) = f_j(\mathbf{k})$ $(i \neq j)$ also along a line. Thus, in general we expect that the condition $f_1(\mathbf{k}) = f_2(\mathbf{k}) = f_3(\mathbf{k}) = 0$ is met in a single point $\mathbf{k}_0$—a *Weyl point* (cf. Fig. 3.1). If we consider a ball around a Weyl point, the corresponding synthetic flux of this two-dimensional surface is $\gamma = \pm 2\pi$, i.e., the Chern number $C = \pm 1$. If we were to enlarge this ball to cover the whole Brillouin zone it is equivalent to a single point due to the periodic boundary conditions of the Brillouin zone, and thus, we must have in this case $\gamma = 0$ (as a result, single Weyl points are only allowed in the continuum). This implies that Weyl points come in pairs with opposite Chern numbers. The pair of Weyl points forms a *Weyl node*, and the only way to annihilate it is by merging the two points. It is possible to have Weyl points with $C = \pm 2$, but as we have seen in Sect. 3.2.1, the dispersions will not be linear in all directions in such a case—like for the case of a glancing intersection/Renner–Teller CI.

The Dirac equation in d dimensions is written compactly as $(i\gamma^\mu \partial_\mu - m)\psi = 0$, where γ^μ, with $\mu = 0, 1, \ldots, d$, are called *gamma matrices* and obey $\{\gamma^\mu, \gamma^\nu\} = 0$ if $\mu \neq \nu$ and otherwise $(\gamma^0)^2 = -(\gamma^i)^2 = 1$, $i = 1, 2, \ldots, d$. The smallest size of these matrices is given by 2^{k+1}, where k is determined by the dimensions according to $d = 2k + 1$ and $2k + 2$. Thus, in one and two dimensions they are 2×2 matrices, while they are 4×4 matrices in three dimensions. In odd spatial dimensions, Weyl pointed out that the Dirac equation can be written in a simple form. In one dimensions and for $m = 0$, which is the case we are interested in, we may take $\gamma^0 = \sigma_z$ and $\gamma^1 = i\sigma_y$, resulting in

$$i\partial_t \psi = \gamma^0 \gamma^1 \hat{p}_x \psi \equiv \gamma^5 \hat{p}_x \psi = \sigma_x \hat{p}_x \psi, \tag{4.89}$$

or in the eigenbasis of σ_x, we simply get $i\partial_t \psi_\pm = \pm p\psi_\pm$. Note that we have defined $\gamma^5 = \gamma^0 \gamma^1$.

In three dimensions, relevant for us, the gamma matrices are 4×4 and can be represented as $\gamma^0 = \mathbb{I} \otimes \tau_x$, $\gamma^i = \sigma_i \otimes \tau_y$, and $\gamma^5 = -\mathbb{I} \otimes \tau_z$, where the τ_i are again Pauli matrices. As in one dimensions, in the eigenbasis of γ^5, i.e., $\gamma^5 \psi_\pm = \pm \psi_\pm$, the Dirac equation becomes

$$i\partial_t \psi_\pm = \mp \hat{\mathbf{p}} \cdot \boldsymbol{\sigma} \psi_\pm \tag{4.90}$$

with eigenvalues $\pm p = \pm |\mathbf{p}|$. This is the characteristic of *Weyl fermions*; they propagate parallel or anti-parallel with respect to their spin.

We note that in the vicinity of the Weyl point, the physics could in some cases (see below) be described by a relativistic equation of the form (4.90). To get a better feeling for the dispersion consider the Bloch Hamiltonian

$$h_{\text{eff}}(\mathbf{k}) = v\tau_x(\mathbf{k}\cdot\boldsymbol{\sigma}) + m\tau_z + b_z\sigma_z + b_x\tau_z\sigma_x$$

$$= \begin{bmatrix} m + b_z\sigma_z + b_x\sigma_x & v\mathbf{k}\cdot\boldsymbol{\sigma} \\ v\mathbf{k}\cdot\boldsymbol{\sigma} & -m + b_z\sigma_z - b_x\sigma_x \end{bmatrix} \quad (4.91)$$

that turns out to show a very rich variety of dispersions. Here, m is the mass and we have included two additional terms representing magnetic fields in the x- and z-directions with amplitudes b_x and b_z, respectively. Depending on the values of the three parameters m, b_x, and b_z, the eigenvalues display qualitatively different behaviour. In Fig. 4.11, we present examples of the four bands for $k_x = 0$ and by either letting all parameters $m = b_1 = b_2 = 0$ or by keeping one of them non-zero. In Fig. 4.11b, when all three parameters are zero, we regain a three-dimensional Dirac CI, or Weyl CI, and the bands are doubly degenerate. With a non-zero mass m the spin degeneracy of the bands survives, but expectedly the Weyl

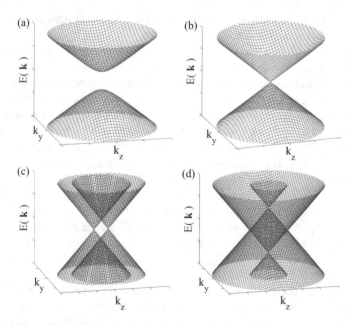

Fig. 4.11 Four examples of dispersions for the Bloch Hamiltonian (4.91) when keeping $k_x = 0$. In all examples the energies are symmetric around the zero energy $E = 0$. In (**a**) we have $(m, b_x, b_z) = (\text{constant}, 0, 0)$ implying that a gap opens up, but the doubly degeneracy is preserved such that both surfaces represent two identical energy dispersions. For a vanishing mass, $(m, b_x, b_z) = (0, 0, 0)$, a Dirac-like CI appears at the origin (**b**). If a time-reversal breaking magnetic field is added the degeneracy is lifted and we find the dispersions of Weyl semimetals. In (**c**) the field is in the x-direction, $(m, b_x, b_z) = (0, \text{constant}, 0)$, and the two Weyl CIs split in the horizontal z-direction. The two Weyl CIs have opposite chirality with corresponding Chern numbers $C = \pm 1$. With the magnetic field in the z-direction, $(m, b_x, b_z) = (0, 0, \text{constant})$, the Weyl CIs split vertically (**d**). As a result a ring-shaped Weyl 'line node' CI appears for zero energy

CI is split as shown in Fig. 4.11a. The two magnetic fields have the effect of lifting the degeneracy; the field in the x-direction ($b_x \neq 0$) displaces the two Weyl CIs horizontally, while a field in the z-direction ($b_z \neq 0$) shifts them energetically in the vertical direction. Figure 4.11c represents a Weyl semimetal with two Weyl points of opposite Chern numbers, $C = \pm 1$. In Fig. 4.11d, where $b_z \neq 0$, there is a Weyl 'line node' at zero energy $E = 0$. Each point along the line has its partner on the opposite side of the ring, in agreement with the observation that Weyl points must come in pairs.

Figure 4.11c serves as our example of dispersion for a Weyl semimetal; a pair of Weyl CIs appear in the Brillouin zone. This is due to a magnetic field that breaks time-reversal symmetry. In fact, to have a Weyl semimetal we need to break either time-reversal symmetry, inversion symmetry, or both. When time-reversal symmetry is broken, while inversion symmetry still holds, we have at least two Weyl CIs at the same energy. In the reversed situation of preserved time-reversal symmetry but broken inversion symmetry, we instead have a minimum of four Weyl CIs. This can be understood from the observation that if having a Weyl CI at $\mathbf{k}_0$ with a certain chirality, then time inversion symmetry implies another Weyl CI at $-\mathbf{k}_0$ with the same chirality. So we must have another two Weyl CIs with opposite chirality in order to get a net zero chirality.

In Fig. 4.11c with two Weyl CIs at $E = 0$, the system is only a true semimetal if the Fermi energy $E_F = 0$. As the Fermi energy departs from zero the Fermi surface is no longer points but two approximate spheres, and, as argued above, we can define the corresponding flux through such surfaces which will be quantised to an integer. The Weyl CI acts as a synthetic magnetic monopole. However, if the Fermi energy is further changed away from $E = 0$, we reach a critical point where the two spheres merge into one surface. Such a change in the Fermi surface is usually termed a *Lifshitz transition*. There is another way in which the character of the Fermi surface may qualitatively change, in *type-II* Weyl semimetals.

We end this section with an interesting observation. First, note that a Weyl CI is robust to perturbations. We know that a mass term splits a Dirac CI, but the same kind of splitting is not allowed for a Weyl CI. If we have the Hamiltonian $h(\mathbf{k}) = v\mathbf{k} \cdot \boldsymbol{\sigma}$, by adding a term say $\delta\sigma_z$ the only effect is to shift the Weyl CI from $(k_x, k_y, k_z) = (0, 0, 0)$ to $(k_x, k_y, k_z) = (0, 0, -\delta)$. More interesting is if the Hamiltonian is modified as $h(\mathbf{k}) = v\mathbf{k} \cdot \boldsymbol{\sigma} + v_0 k_z \mathbb{I}$, i.e., a momentum dependent constant $v_0 k_z$ has been included. The effect of such a term is to tilt the Weyl CI. Taking $E_F = 0$, a small tilt does not qualitatively alter the physical properties. But as soon as $|v_0| > |v|$ the CI 'tips over' as depicted in Fig. 4.12, and the system is then called a type-II Weyl semimetal. The Fermi surface is no longer closed but open, even though it is still a semimetal since the gap closes in a single point.

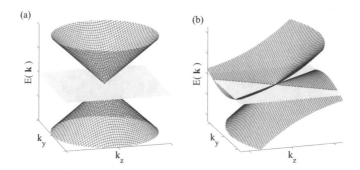

Fig. 4.12 A type-I (**a**) and a type-II (**b**) Weyl semimetal, where we imagine $k_x = 0$ and the light grey plane represents the $E_F = 0$ Fermi energy. For a type-I semimetal, this Fermi surface is point-like, while for a type-II it is an open line. The transition occurs when the tilting of the Weyl CI becomes sufficiently large so that it tips over

References

1. Kohn, W.: Analytic properties of Bloch waves and Wannier functions. Phys. Rev. **115**, 809 (1959)
2. Brouder, C., Panati, G., Calandra, M., Mourougane, C., Marzari, N.: Exponential localization of Wannier functions in insulators. Phys. Rev. Lett. **98**, 046402 (2007)
3. Bernevig, B.A.: Topological Insulators and Topological Superconductors. Princeton University Press, Princeton/Oxford (2013)
4. Altland, A., Zirnbauer, M.R.: Nonstandard symmetry classes in mesoscopic normal-superconducting hybrid structures. Phys. Rev. B **55**, 1142 (1997)
5. Zak, J.: Berry's phase for energy bands in solids. Phys. Rev. Lett. **62**, 2747 (1989)
6. Thouless, D.J., Kohmoto, M., Nightingale, M.P., den Nijs, M.: Quantized Hall conductance in a two-dimensional periodic potential. Phys. Rev. Lett. **49**, 405 (1982)
7. Stanescu, T.D.: Introduction to Topological Quantum Matter and Computing. CRC Press, New York (2017)
8. Kane, C.L., Mele, E.J.: Quantum spin Hall effect in graphene. Phys. Rev. Lett. **95**, 226801 (2005)
9. Kane, C.L., Mele, E.J.: Z_2 Topological order and the quantum spin Hall effect. Phys. Rev. Lett. **95**, 146802 (2005)
10. Manchon, A., Koo, H.C., Nitta, J., Frolov, S.M., Duine, R.A.: New perspectives for Rashba spin-orbit coupling. Nature Mat. **14**, 871 (2015)
11. Sinova, J., Culcer, D., Niu, Q., Sinitsyn, N.A., Jungwirth, T., MacDonald, A.H.: Universal intrinsic spin Hall effect. Phys. Rev. Lett. **92**, 126603 (2004)
12. Bostwick, A., Ohta, T., Seyller, T., Horn, K., Rotenberg, E.: Quasiparticle dynamics in graphene. Nature Phys. **3**, 36 (2007)
13. Tarruell, L., Greif, D., Uehlinger, T., Jotzu, G., Esslinger, T.: Creating, moving and merging Dirac points with a Fermi gas in a tunable honeycomb lattice. Nature **483**, 302 (2012)
14. Haldane, F.D.M.: Model for a quantum Hall effect without Landau levels: condensed-matter realization of the "Parity Anomaly". Phys. Rev. Lett. **61**, 2015 (1988)
15. Jotzu, G., Messer, M., Desbuquois, R., Lebrat, M., Uehlinger, T., Greif, D., Esslinger, T.: Experimental realization of the topological Haldane model with ultracold fermions. Nature **515**, 237 (2014)
16. Armitage, N.P., Mele, E.J., Vishwanath, A.: Weyl and Dirac semimetals in three-dimensional solids. Rev. Mod. Phys. **90**, 015001 (2018)

Chapter 5
Conical Intersections in Cold Atom Physics

Abstract In this chapter, we discuss how conical intersections can appear in systems of cold atoms. We in particular study atoms which are interacting with light with emphasis on continuum systems. A key ingredient in this respect is the concept of synthetic magnetism and corresponding gauge potentials, which can appear from the adiabatic motion of the atoms in combination with light coupled to the internal level structure of the atoms. This leads us to the concept of spin–orbit coupling and conical intersections (CIs). We discuss the role of collisions between the atoms in combination with spin–orbit coupling, and make a connection to quasi-relativistic dynamics with cold atoms.

5.1 Introduction

The combination of light coupled to a gas of cold atoms has provided us with a remarkable scenario where it is possible to study in experiments macroscopic quantum dynamics. Light plays a key role here. It is a versatile tool which can be used to manipulate macroscopic objects but also individual atoms. With the invention of the laser in the 1950s it became possible to accurately probe the internal structure of atoms but also to exert forces on the atoms and thereby control their dynamics.

The laser also made it possible to laser cool ensembles of atoms, and in combination with magnetic trapping, the first degenerate quantum gases were created. Bosonic atoms but also fermionic atoms are now routinely prepared in laboratories all over the world, where temperatures of the order of micro Kelvin and below are reached. The resulting Bose–Einstein condensates (BECs) and degenerate Fermi gases have proven to be remarkably useful tools to study exotic phenomena known from almost any branch of physics.

These quantum gases have helped us understand many concepts from traditional condensed matter physics, such as the superfluid to Mott phase transition and the Bardeen–Cooper–Schrieffer crossover in degenerate Fermi gases [1–5]. Surprisingly also phenomena known from high-energy physics can be emulated in these ultracold settings. For instance, Higgs modes have been observed and studied

© Springer Nature Switzerland AG 2020
J. Larson et al., *Conical Intersections in Physics*, Lecture Notes in Physics 965,
https://doi.org/10.1007/978-3-030-34882-3_5

[6, 7], confinement mechanisms for quarks [8], axion electrodynamics [9], and unconventional colour superconductivity [10] have also been proposed. The rather unique access to experimental parameters such as trap geometry, particle number and density of the cloud, and even the collisional interaction strength between the atoms using Feshbach resonances makes it possible to probe many different physical phenomena.

Atomic quantum gases typically have no charge, and therefore, they are not affected by external electromagnetic fields the way electrons are. The atom–light coupling does, however, allow for the creation of versatile gauge potentials, which allows for experimental access to many new phenomena related to magnetism at the quantum level. With this technology, atoms can be subject to static Abelian and non-Abelian [11] gauge fields, which in practice means that synthetic electric and magnetic fields can be experimentally controlled with lasers. The non-Abelian gauge fields can reproduce Rashba-type spin–orbit couplings, but also emulate a variety of properties encountered in the context of high-energy physics. Mimicking magnetic and spin–orbit effects in ultracold quantum gases also makes it possible to envisage quantum simulators of new kinds of exotic quantum matter [12] and realise Feynman's vision [13] for constructing physical quantum emulators of systems or situations that are computationally intractable.

Synthetic gauge fields can also be created for atoms which are confined by an optical lattice. This in turn opens up a natural link to quantum simulations of condensed matter phenomena. In the optical lattice the synthetic magnetic field can be created using laser-induced tunnelling between the lattice sites, or by modulating lattice parameters such that an effective Hamiltonian emulates the presence of a magnetic field. This has opened up the possibility to seriously consider proposals for simulating exotic condensed matter models and also realising strongly correlated systems [14].

In this chapter, we first briefly review how atoms interact with light and how they can be manipulated by light. To illustrate synthetic magnetism, we consider the simplest possible scenario where we use a two-level atom illuminated by a laser field, for which a gauge potential appear for the atoms. We then discuss a more general situation where the light–matter interaction provides us with the mechanisms for synthetic spin–orbit coupling. As an illustration we study the emerging quasi-relativistic dynamics that can be emulated in such situations where the atoms are governed by a Dirac equation. In the final sections we present a theoretical description of a Bose–Einstein condensate and discuss what role interactions play in the context of CIs.

5.2   Light–Matter Interactions and Optical Forces

The interaction between light and the atoms is a key ingredient if we want to use cold atoms for emulating synthetic gauge potentials and spin–orbit coupling, which consequently gives us access to the CIs in the dispersion relation for the atoms. In order to show how to manipulate atoms with light we start from classical electromagnetism, where it is known that light exerts a force on classical dipoles. This force depends on the gradient of the amplitude and phase of the light [15],

$$m\ddot{\mathbf{R}} = \mathbf{d} \cdot (\nabla \mathbf{E}) = \mathbf{d} \cdot \boldsymbol{\epsilon}(\nabla \xi + \xi \nabla \theta)e^{i(\theta(\mathbf{R})+\omega t)}, \tag{5.1}$$

where m is the mass of the dipole, $\mathbf{d}$ is the dipole moment, and $\mathbf{E}(\mathbf{R}, t) = \boldsymbol{\epsilon}\xi(\mathbf{R})e^{i(\omega t+\theta(\mathbf{R}))}$ is the electric field with amplitude ξ, polarisation $\boldsymbol{\epsilon}$, phase θ, and frequency ω.

The two-level atom, shown in Fig. 5.1, can also be seen as a dipole. We describe in the following the atom quantum-mechanically, but allow the light field to be classical. The Hamiltonian for such a situation is

$$\hat{H} = \frac{\hat{\mathbf{P}}^2}{2m} + \hat{H}_0 - \hat{\mathbf{d}} \cdot \mathbf{E}(\mathbf{R}, t), \tag{5.2}$$

where $\hat{\mathbf{P}}^2/2m$ is the kinetic energy associated with the centre of mass motion of the atom, $\hat{H}_0$ is the Hamiltonian for the unperturbed internal motion, and $\hat{\mathbf{d}} \cdot \mathbf{E}(\mathbf{R}, t)$ is the interaction between the atom and the light field, which is based on the dipole approximation. By using Ehrenfest's theorem with the Hamiltonian in (5.2), the force can be expressed as

$$F = m\ddot{\mathbf{r}} = \langle \nabla(\hat{\mathbf{d}} \cdot \mathbf{E}) \rangle = \langle \hat{\mathbf{d}} \cdot \boldsymbol{\epsilon} \rangle \nabla \xi(\mathbf{r}, t) \tag{5.3}$$

with $\mathbf{r} = \langle \mathbf{R} \rangle$ and $\xi(\mathbf{r}, t) = \xi(\mathbf{r})e^{i(\omega t+\theta(\mathbf{r}))}$. In the right-hand side above we have assumed that the force is uniform across the atomic wave packet.

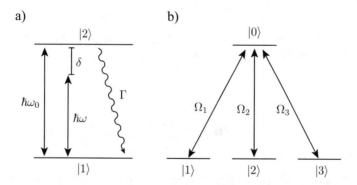

Fig. 5.1 (a) The two-level atom with its coupled two states $|1\rangle$ and $|2\rangle$, where the energy difference is $\hbar\omega_0$. The frequency of the external laser that couples the two levels is ω, with detuning $\delta = \omega - \omega_0$. The decay rate of the excited state $|2\rangle$ is Γ. (b) The tripod configuration of the atom–light coupling in which the atomic ground states $|1\rangle$, $|2\rangle$, $|3\rangle$ are coupled to the excited state $|0\rangle$ via the Rabi frequencies Ω_1, Ω_2, Ω_3 of the laser fields

A two-level atom driven by a laser is the drosophila of quantum optics and light–matter interaction. In order to obtain an expression for the force acting on the atom we need to calculate the response of the atom to the light, i.e. the susceptibility or polarisation, $\langle \hat{\mathbf{d}} \cdot \boldsymbol{\epsilon} \rangle$. For this we assume a monochromatic field of the form

$$\xi(\mathbf{r}, t) = \frac{1}{2} E(\mathbf{r}) e^{i(\theta(\mathbf{r}) + \omega t)}, \tag{5.4}$$

where $E(\mathbf{r})$ is the amplitude, ω the laser frequency, and θ a space- dependent phase factor. The corresponding Schrödinger equation then gives us two coupled equations for the probability amplitudes C_1 and C_2 for the atom to be found in state $|1\rangle$ and $|2\rangle$, respectively. We can choose a rotating frame with

$$C_1 = D_1 e^{i\frac{1}{2}(\theta + \delta t)},$$
$$C_2 = D_2 e^{-i\frac{1}{2}(\theta + \delta t)}, \tag{5.5}$$

which results in the equations

$$i\dot{D}_1 = \frac{1}{2}(\dot{\theta} + \delta) D_1 - \frac{\Omega}{2} D_2,$$
$$i\dot{D}_2 = -\frac{1}{2}(\dot{\theta} + \delta) D_2 - \frac{\Omega}{2} D_1. \tag{5.6}$$

In the equations for D_1 and D_2, we have introduced the detuning $\delta = \omega - \omega_0$ and used the rotating wave approximation (RWA), where rapidly oscillating terms are neglected. The dipole moment d for the transition between state 1 and 2 is given by $d = \langle 1|\hat{\mathbf{d}} \cdot \boldsymbol{\epsilon}|1\rangle$ and the Rabi frequency is defined by

$$\Omega = \frac{d E(t)}{\hbar}. \tag{5.7}$$

It is convenient to describe the system using the density matrix, which is defined as $\rho_{nm} = C_n C_m^*$ or $\sigma_{nm} = D_n D_m^*$ with

$$\rho_{11} = \sigma_{11},$$
$$\rho_{22} = \sigma_{22},$$
$$\rho_{12} = \sigma_{12} e^{i(\theta + \omega t)},$$
$$\rho_{21} = \sigma_{21} e^{-i(\theta + \omega t)}. \tag{5.8}$$

From (5.6), we see that the matrix elements of the density matrix obey

$$\dot{\sigma}_{11} = -\frac{i}{2}\Omega(\sigma_{12} - \sigma_{21}) + \Gamma\sigma_{22},$$
$$\dot{\sigma}_{22} = \frac{i}{2}\Omega(\sigma_{12} - \sigma_{21}) - \Gamma\sigma_{22},$$

$$\dot{\sigma}_{12} = -i(\delta + \dot{\theta})\sigma_{12} + \frac{i}{2}\Omega(\sigma_{22} - \sigma_{11}) - \frac{1}{2}\Gamma\sigma_{12}, \tag{5.9}$$

where we have introduced the spontaneous emission rate Γ which describes decay processes.

The density matrix allows us to calculate expectation values. For our purpose we need the expectation value of the dipole moment, which is given by

$$\langle \hat{\mathbf{d}} \cdot \boldsymbol{\epsilon} \rangle = d(\rho_{12} + \rho_{21}) = d(\sigma_{12}e^{i(\theta + \omega t)} + \sigma_{21}e^{-i(\theta + \omega t)}). \tag{5.10}$$

With this expression, together with the RWA, we get from (5.3) and (5.4) the force

$$\mathbf{F} = \frac{d}{2}(\sigma_{12} + \sigma_{21} - i(\sigma_{12} - \sigma_{21})) = \frac{\hbar}{2}(U\nabla\Omega + V\Omega\nabla\theta), \tag{5.11}$$

where we have introduced the notation $U = \sigma_{12} + \sigma_{21}$ and $V = i(\sigma_{12} - \sigma_{21})$, and used the fact that $\dot{\theta} = \nabla\theta(\mathbf{r}) \cdot \dot{\mathbf{r}}$. If the atomic motion is slow, the phase $\dot{\theta}$ of the atomic state does not change much during the lifetime $1/\Gamma$ of the excited state. We can therefore use the steady state solution of the density matrix and put the time derivatives of the left-hand side equal to zero in (5.9). The solutions of the corresponding U and V are

$$U = \frac{\delta}{\Omega}\frac{s}{s+1},$$

$$V = \frac{\Gamma}{2\Omega}\frac{s}{s+1}, \tag{5.12}$$

where s is the saturation parameter

$$s = \frac{\Omega^2/2}{(\delta + \dot{\theta})^2 + \Gamma^2/4}. \tag{5.13}$$

The resulting force acting on the atom has two parts which are referred to the dipole force and the radiation force, respectively,

$$\mathbf{F} = \mathbf{F}_{\text{dip}} + \mathbf{F}_{\text{pr}}, \tag{5.14}$$

where

$$\mathbf{F}_{\text{dip}} = -\frac{\hbar(\delta + \dot{\theta})}{2}\frac{\nabla s}{s+1},$$

$$\mathbf{F}_{\text{pr}} = -\frac{\hbar\Gamma}{2}\frac{s}{s+1}\nabla\theta. \tag{5.15}$$

For plane waves the radiation force $\mathbf{F}_{pr}$ is proportional to the wave vector $\mathbf{k} = \nabla\theta$. If we, however, want to trap the atom, the dipole force $\mathbf{F}_{dip}$ is more important. The force $\mathbf{F}_{dip}$ is determined by the intensity of the laser field. If $s \ll 1$ and $|\delta| \ll \Omega$, we get the corresponding potential using $\mathbf{F}_{dip} = \nabla W$, where

$$W = \frac{\hbar\Omega^2}{4\delta} = \frac{d^2 E^2}{4\delta\hbar}. \tag{5.16}$$

From this expression we see that if the intensity of the light is inhomogeneous we obtain a non-zero force whose direction depends on the sign of the detuning. For a focused Gaussian beam this means that the atoms are attracted to the high intensity if the laser is red detuned ($\delta < 0$), i.e. the atoms are high field seekers. On the other hand, if the laser is blue detuned ($\delta > 0$), the atoms are low field seekers and are repelled from the centre of the beam.

We see that the dipole force $\mathbf{F}_{dip}$ can therefore be used as a versatile tool for controlling the atoms. In particular, the dipole potential created by two pairs of laser beams counter-propagating at a right angle provides a square optical lattice. More exotic lattices can also be created. For instance, hexagonal or triangular optical lattices can be made using three laser beams propagating at 120°, or an edge-centred honeycomb lattice, with a five band structure, can be created by choosing six lasers as

$$\begin{aligned}
\mathbf{E}_1 &= E(0, 1)e^{ik\mathbf{x}\cdot\mathbf{a}_1}, \\
\mathbf{E}_2 &= E(\sqrt{3}/2, 1/2)e^{-ik\mathbf{x}\cdot\mathbf{a}_2}, \\
\mathbf{E}_3 &= E(\sqrt{3}/2, -1/2)e^{ik\mathbf{x}\cdot\mathbf{a}_3}, \\
\mathbf{E}_4 &= E(0, -1)e^{-ik\mathbf{x}\cdot\mathbf{a}_1}, \\
\mathbf{E}_5 &= E(-\sqrt{3}/2, -1/2)e^{ik\mathbf{x}\cdot\mathbf{a}_2}, \\
\mathbf{E}_6 &= E(-\sqrt{3}/2, 1/2)e^{-ik\mathbf{x}\cdot\mathbf{a}_3}, \tag{5.17}
\end{aligned}$$

where $\mathbf{a}_1 = (1, 0)$, $\mathbf{a}_2 = (-1/2, \sqrt{3}/2)$, $\mathbf{a}_3 = (-1/2, -\sqrt{3}/2)$ are the three nearest-neighbour vectors of the honeycomb structure and k is the wave vector of the lasers, see Fig. 5.2. The intensity profile $I(x, y) = |\mathbf{E}_{tot}(x, y)|^2$ from the total electric field $\mathbf{E}_{tot} = \sum_i \mathbf{E}_i$ produces a potential landscape as shown in Fig. 5.2.

These optical lattices provide a natural connection to condensed matter physics, where it is now possible to design the underlying lattice structure in a controlled way and effectively emulate any material structures where the atoms trapped by the optical lattice take the role of electrons. We can therefore apply the techniques described in Chap. 4 to identify CIs and other aspects such as transport properties of the particles. In this chapter, we concentrate on the continuum regime where the cold atoms are trapped by a confining potential, which is not of a lattice structure. The light–matter interaction, however, provides a mechanism for creating synthetic magnetic fields and spin–orbit coupling, which in turn can result in a spectrum with

Fig. 5.2 With six lasers chosen as in (5.17) the resulting optical potential forms an edge-centred honeycomb lattice for the cold atoms. Red indicates high intensity and blue low intensity

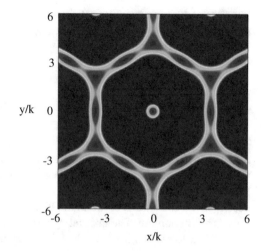

CIs and quasi-relativistic dynamics. Before we get there we need to discuss how to theoretically describe interacting cold atoms and in particular the BEC.

5.3   Adiabatic Dynamics and Synthetic Gauge Potentials

A bosonic quantum gas in the continuum regime is governed by a Schrödinger equation. In the mean-field regime and in the presence of collisions between the atoms, this results in the Gross–Pitaevskii equation, which is described in Sect. 5.5. Our goal is to identify CIs in the continuum regime, in contrast to the discrete system, which is described by a lattice model. For this we need to induce gauge potentials into the system such that we emulate an effective magnetic field and in particular spin–orbit coupling, which gives access to dispersion relations with Dirac cones. The underlying mechanism for achieving this relies on the concept of synthetic gauge potentials and the adiabatic dynamics of the atoms.

It is instructive to start from the simplest possible situation based on a two-level system illuminated by an incident laser beam, in order to illustrate the emergence of synthetic gauge potentials for neutral atoms. It is important to note that, since the atoms have no charge, they do not behave as electrons would do in a real magnetic field. In other words, there is no Lorenz force for instance. We therefore need to induce a synthetic magnetic field, such that the atoms behave *as if* they had a charge. For this we rely on light–matter interaction. There are in fact many different ways one can create synthetic magnetic fields. Here, we concentrate on one particular method, which relies on geometric potentials. For other methods such as using a rotating frame of reference, Raman transitions or Floquet methods, we refer the reader to review articles, see, for instance, [14]. As we show next, the two-level example can be generalised such that the gauge potential becomes a matrix, see

Fig. 5.1b, which consequently defines a pseudo-spin and gives us access to spin–orbit coupling. Before we get there, we must discuss the adiabatic principle in the light–matter systems.

5.3.1 The Adiabatic Principle and Dressed States

The concept of a Berry phase can be translated to the situation where an atom is moving slowly in a monochromatic laser field. In this situation, we have two degrees of freedom to consider: the centre of mass motion, which is described by the position $\mathbf{r}$ and the momentum $\mathbf{p} = i\hbar\nabla$ of the atom, and the internal dynamics, which describes the transitions between the discrete energy levels of the atom due to the light–matter coupling. The internal dynamics in turn is described by a Hamiltonian $\hat{H}_{int}(\mathbf{r})$. The total Hamiltonian, which includes the kinetic part and the light coupling to the internal structure of the atoms, consequently becomes

$$\hat{H}_{tot} = \frac{\hat{\mathbf{p}}^2}{2M} + \hat{H}_{int}. \tag{5.18}$$

As far as the internal Hamiltonian $\hat{H}_{int}(\mathbf{r})$ is concerned, $\mathbf{r}$ is just an external parameter, and we can solve for the eigenstates of $\hat{H}_{int}(\mathbf{r})$. These states are the *dressed states* that satisfy

$$\hat{H}_{int}(\mathbf{r})|\psi_n(\mathbf{r})\rangle = \epsilon_n(\mathbf{r})|\psi_n(\mathbf{r})\rangle. \tag{5.19}$$

The dressed states $|\psi_n(\mathbf{r})\rangle$ form a basis for the internal Hilbert space of the atom where n labels the discrete levels. We therefore can write the full quantum state of the system as

$$|\Psi(\mathbf{r}, t)\rangle = \sum_n \varphi_n(\mathbf{r}, t)|\psi_n(\mathbf{r})\rangle. \tag{5.20}$$

We are interested in the orbital magnetism of the atom, and therefore the centre of mass motion of the atom. From (5.20), we see that this is captured by the complex function $\varphi(\mathbf{r}, t)$, being the probability amplitude to find the atom at position $\mathbf{r}$ in the internal state $|\psi_n(\mathbf{r})\rangle$. Next, we need to incorporate the adiabatic principle. We assume the atom is prepared in one of the internal states, say $|\psi_q(\mathbf{r})\rangle$. We suppose the atom moves slowly enough so that it remains in this state. In other words, we neglect any coupling to the other internal states of the atoms. This is the adiabatic assumption. We can obtain an equation of motion for the corresponding $\varphi(\mathbf{r}, t)$ by starting from the Schrödinger equation

$$i\hbar\frac{\partial}{\partial t}|\Psi(\mathbf{r}, t)\rangle = \hat{H}_{tot}|\Psi(\mathbf{r}, t)\rangle = \left[\frac{\hbar^2}{2M}\nabla^2 + \hat{H}_{int}\right]|\Psi(\mathbf{r}, t)\rangle. \tag{5.21}$$

By inserting the expansion (5.20) into (5.21), and projecting onto the specific dressed state $|\psi_q(\mathbf{r})\rangle$, which we assume the particle remains in, we obtain an equation of motion of the form

$$i\hbar\frac{\partial}{\partial t}\varphi_q(\mathbf{r},t) = \left[\frac{(\mathbf{p} - \mathbf{A}_q(\mathbf{r}))^2}{2M} + \epsilon_q(\mathbf{r}) + \mathscr{V}_q(\mathbf{r})\right]\varphi_q(\mathbf{r},t). \tag{5.22}$$

This has the structure of a Schrödinger equation for a charged particle which is subject to a gauge potential of the form

$$\mathbf{A}_q(\mathbf{r},t) = i\hbar\langle\psi_q|\nabla\psi_q\rangle, \tag{5.23}$$

which coincides with (2.55) up to a factor $\hbar$. This is precisely the form of the synthetic gauge connection. There are two more terms in (5.22). The first one is the energy $\epsilon_q(\mathbf{r})$ of the dressed state which the atom is prepared in, and the second one is the scalar potential (cf. (2.85))

$$\mathscr{V}_q(\mathbf{r}) = \frac{\hbar^2}{2M}\sum_{n\neq q}|\langle\nabla\psi_q|\psi_n\rangle|^2. \tag{5.24}$$

This scalar potential can be understood as the energy associated with the micro motion of the atom which stems from the virtual transitions between the dressed states $|\psi_q\rangle$ and all the other dressed states ψ_n with $n \neq q$.

5.3.2 A Pedagogical Example: The Two-Level System

Perhaps the simplest example that illustrates how synthetic gauge potentials come about is based on a single particle, which we describe quantum-mechanically. The particle has two internal energy levels $|g\rangle$ and $|e\rangle$, which are coupled by classical light. The main idea is to solve exactly the internal dynamics and then allow the particle to also have kinetic energy. As we see next, the resulting Schrödinger equation for the centre of mass motion acquires a gauge potential that depends on the parameters of the incident light.

We start by describing the internal degree of freedom of the particle by $\{|g\rangle, |e\rangle\}$ that forms a basis of the two-dimensional Hilbert space. For an atom, these states are the electronic ground and excited states, where we neglect any spontaneous emission.

We assume that the particle moves in some space-dependent external field that couple $|g\rangle$ and $|e\rangle$. This can be the field of an optical laser, microwave fields, static electric or magnetic fields which act on the electric or magnetic dipole moment of the particle. The Hamiltonian that describes this coupling can be written in the matrix form

$$H_{int} = \frac{\hbar}{2} \begin{pmatrix} \Delta & \kappa^* \\ \kappa & -\Delta \end{pmatrix} \tag{5.25}$$

where now the Rabi frequency κ and detuning Δ are allowed to depend on the atom's centre of mass position, see Fig. 5.1. At this point it is instructive to parametrise the Hamiltonian in (5.25) by defining the generalised Rabi frequency Ω, the mixing angle θ, and the phase ϕ with

$$\Omega = \sqrt{\Delta^2 + |\kappa|^2},$$

$$\tan \theta = \frac{|\kappa|}{\Delta},$$

$$\kappa = |\kappa| e^{i\phi}, \tag{5.26}$$

which results in the atom-laser coupling

$$H_{int} = \frac{\hbar \Omega}{2} \begin{pmatrix} \cos \theta & e^{-i\phi} \sin \theta \\ e^{i\phi} \sin \theta & -\cos \theta \end{pmatrix}. \tag{5.27}$$

It is of course a matter of taste which description one uses, but the mixing angle description often tends to simplify the algebra.

The eigenvalues of the Hamiltonian in (5.27) are $\pm \hbar \Omega / 2$ with the eigenstates

$$|\chi_+\rangle = \begin{pmatrix} \cos(\theta/2) \\ e^{i\phi} \sin(\theta/2) \end{pmatrix} \tag{5.28}$$

$$|\chi_-\rangle = \begin{pmatrix} -e^{-i\phi} \sin(\theta/2) \\ \cos(\theta/2) \end{pmatrix}. \tag{5.29}$$

If the state of the atom follows adiabatically one of the dressed states $|\chi_\pm\rangle$, we obtain from (5.23) the gauge potential

$$\mathbf{A}_\pm = \pm \frac{\hbar}{2}(1 - \cos \theta)\nabla\phi. \tag{5.30}$$

The corresponding synthetic magnetic field can then readily be calculated as

$$\mathbf{B}_\pm = \nabla \times \mathbf{A}_\pm = \pm \frac{\hbar}{2}\nabla(\cos \theta) \times \nabla\phi, \tag{5.31}$$

with the scalar potential in (5.24) given by

$$\mathcal{V}_\pm(\mathbf{r}) = \frac{\hbar^2}{8M} \left[(\nabla\theta)^2 + \sin^2\theta (\nabla\phi)^2 \right].$$ (5.32)

From the expressions for the gauge potential and the magnetic field we see that in order to have non-zero magnetic fields we must have non-zero gradients of the phase ϕ and the mixing angle θ. A space-dependent mixing angle can be achieved either by having a non-zero gradient of the laser intensity or a non-zero gradient of the detuning. The potentials $\mathbf{A}_\pm$, $\mathbf{B}_\pm$, and $\mathcal{V}_\pm$ are geometrical in nature. This can be seen from the expressions (5.30), (5.31), and (5.32), which depend only on the spatial variation of the angles θ and ϕ, but not on the strength Ω of the coupling.

5.4 Spin–Orbit Coupling and Non-Abelian Phenomena

The two-level scheme discussed in the previous section can also be generalised to multi-level systems, see Fig. 5.1b. As we see next, this results in a gauge potential which is a matrix. This in turn gives us access to spin–orbit coupling, where the internal dynamics of the atom is coupled to the centre of mass motion. A matrix gauge potential also allows for non-Abelian phenomena which can be seen as a generalised spin–orbit coupling, where the vector components of the gauge potential are matrices which do not commute. The non-Abelian Aharonov–Bohm experiment is a simple example, where non-trivial spin dynamics can be seen. There it is found that the spin rotates differently depending on which path is taken, as illustrated in Fig. 5.3.

The idea to create non-Abelian gauge potentials based on adiabatic evolution was first proposed by Wilczek and Zee [11]. In a similar manner to the two-level example in Sect. 5.3, we generalise the situation and consider g levels that form a subspace well separated from the rest of the spectrum. With the eigenvectors of the internal Hamiltonian we form an orthonormal basis $\{|\psi_1(\mathbf{r})\rangle, \ldots, |\psi_g(\mathbf{r})\rangle\}$. If the adiabatic assumption holds we can write a general state as

$$|\Psi(\mathbf{r})\rangle = \sum_{n=1}^{g} \phi_n(\mathbf{r})|\psi_n(\mathbf{r})\rangle.$$ (5.33)

Fig. 5.3 The non-Abelian Aharonov–Bohm effect. A particle with spin is subject to a non-Abelian gauge potential, and can take two different paths P_1 or P_2. For the same initial state, the final state can be different depending on which path was taken

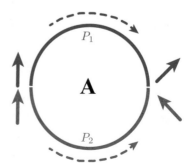

By projecting the full Schrödinger equation onto the g-dimensional subspace spanned by $\{|\psi_1(\mathbf{r})\rangle, \ldots, |\psi_g(\mathbf{r})\rangle\}$ we get g coupled equations of the form

$$i\hbar \frac{\partial}{\partial t} \Phi(\mathbf{r}, t) = \left[\frac{(\mathbf{p} - \mathbf{A}(\mathbf{r}))^2}{2M} + \epsilon(\mathbf{r}) + \mathscr{V}(\mathbf{r}) \right] \Phi(\mathbf{r}, t) \tag{5.34}$$

with

$$\Phi(\mathbf{r}, t) = \begin{pmatrix} \phi_1(\mathbf{r}, t) \\ \vdots \\ \phi_q(\mathbf{r}, t) \end{pmatrix}. \tag{5.35}$$

Here, $\epsilon(\mathbf{r}) = \text{diag}\{\epsilon_1(\mathbf{r}), \ldots, \epsilon_g(\mathbf{r})\}$ and $\mathscr{V}(\mathbf{r})$ are $g \times g$ Hermitian matrices, the latter representing the geometric scalar potential with matrix elements

$$\mathscr{V}_{nm} = \frac{\hbar^2}{2M} \left(\langle \nabla \psi_n | \nabla \psi_m \rangle + \sum_{k=1}^{q} \langle \psi_n | \nabla \psi_k \rangle \langle \psi_k | \nabla \psi_m \rangle \right). \tag{5.36}$$

The gauge potential is now also a $g \times g$ Hermitian matrix whose vector-valued matrix elements are given by

$$\mathbf{A}_{nm}(\mathbf{r}) = i\hbar \langle \psi_n(\mathbf{r}) | \nabla \psi_m(\mathbf{r}) \rangle \tag{5.37}$$

with $n, m \in (1, \ldots, g)$.

The tripod configuration, considered in Sect. 2.3.2 and illustrated in Fig. 5.1b, is one example of an atomic level structure that can give rise to both spin–orbit coupling and non-Abelian effects. The tripod level scheme consists of three atomic levels coupled to a common excited level via laser fields with the Raman frequencies Ω_j.

If the atom–light coupling is on resonance with the transitions, the Hamiltonian of the tripod atom can be written as

$$\hat{H} = \hbar \left(\Omega_1 |e\rangle\langle g_1| + \Omega_2 |e\rangle\langle g_2| + \Omega_3 |e\rangle\langle g_3| \right) + \text{H.c.} = \hbar\Omega \left(|e\rangle\langle B| + \text{H.c.} \right), \tag{5.38}$$

where $\Omega = \sqrt{|\Omega_1|^2 + |\Omega_2|^2 + |\Omega_3|^2}$ is the total Rabi frequency and

$$|B\rangle = \frac{1}{\Omega} \left(\Omega_1^* |g_1\rangle + \Omega_2^* |g_2\rangle + \Omega_3^* |g_3\rangle \right) \tag{5.39}$$

is the atomic bright state, representing a superposition of the atomic ground states directly coupled to the excited state $|e\rangle$. Alternatively we can write the Hamiltonian in matrix form,

$$H_{\text{int}} = \hbar \begin{pmatrix} 0 & \Omega_1^* & \Omega_2^* & \Omega_3^* \\ \Omega_1 & 0 & 0 & 0 \\ \Omega_2 & 0 & 0 & 0 \\ \Omega_3 & 0 & 0 & 0 \end{pmatrix}, \tag{5.40}$$

where we have used the basis $\{|e\rangle, |g_1\rangle, |g_2\rangle, |g_3\rangle\}$. There are two degenerate dark states that are superpositions of the three ground states with no contribution from the excited state, and with eigenvalue zero.

The resulting adiabatic states $|\chi_1\rangle, |\chi_2\rangle$ are degenerate, which means that any linear superposition of them is also a valid eigenstate. We can therefore define a local space-dependent basis change using a unitary transformation $U(\mathbf{r})$. If we change the basis, we also need to transform the gauge potential and its scalar counterpart,

$$\mathbf{A}(\mathbf{r}) \rightarrow U(\mathbf{r})\mathbf{A}(\mathbf{r})U^\dagger(\mathbf{r}) - i\hbar(\nabla U(\mathbf{r}))U^\dagger(\mathbf{r}),$$

$$\mathscr{V}(\mathbf{r}) \rightarrow U(\mathbf{r})\mathscr{V}(\mathbf{r})U^\dagger(\mathbf{r}). \tag{5.41}$$

We therefore see that the act of changing the basis also represents a gauge transformation.

The gauge potential $\mathbf{A}$ is the Wilczek–Zee connection discussed in Sect. 2.3.1. The corresponding magnetic field, which is now also a matrix, is $B_i = \varepsilon_{ikl}F_{kl}$, where

$$F_{kl} = \partial_k A_l - \partial_l A_k - \frac{i}{\hbar}[A_k, A_l] \tag{5.42}$$

is the gauge curvature tensor with ε_{ikl} the Levi-Civita symbol. The non-Abelian nature is captured by the last term in (5.42) and $\frac{1}{2}\varepsilon_{ikl}[A_k, A_l]$, which is not necessarily zero because the matrix vector elements might not commute.

The Hamiltonian $\hat{H}_{\text{int}}$ in (5.38) has two eigenstates $|\chi_1(\mathbf{r})\rangle \equiv |D_1\rangle$ and $|\chi_2(\mathbf{r})\rangle \equiv |D_2\rangle$ known as the dark (or uncoupled) states, which are orthogonal to the bright state $|B\rangle$. The dark states contain no excited state contribution and are characterised by zero eigenenergies ($\hat{H}_{\text{int}}|D_j\rangle = 0$). The bright state $|B\rangle \equiv |B(\mathbf{r})\rangle$ is position dependent due to the position dependence of the amplitude and the phase of the laser Rabi frequencies $\Omega_j \equiv \Omega_j(\mathbf{r})$. Thus the adiabatic elimination of the bright and excited states leads to the adiabatic centre of mass motion of the dark state atoms affected by the vector potential $\mathbf{A}^{(g)}$ (with $g = 2$ since there are two dark states) to be labelled simply by $\mathbf{A}$. The explicit expressions for the vector potential $\mathbf{A}$ and the accompanying geometric scalar potential can readily be calculated using the eigenstates, i.e. the dark states, of the Hamiltonian in (5.40),

$$|D_1\rangle = \sin\phi e^{iS_{31}}|g_1\rangle - \cos\phi e^{iS_{32}}|g_2\rangle,$$

$$|D_2\rangle = \cos\theta\cos\phi e^{iS_{31}}|g_1\rangle + \cos\theta\sin\phi e^{iS_{32}}|g_2\rangle - \sin\theta|g_3\rangle \tag{5.43}$$

with $S_{ij} = S_i - S_j$. Here, we have parametrised the system according to

$$\Omega_1 = \Omega \sin\theta \cos\phi e^{iS_1},$$
$$\Omega_2 = \Omega \sin\theta \sin\phi e^{iS_2},$$
$$\Omega_3 = \Omega \cos\theta e^{iS_3} \tag{5.44}$$

with $\Omega = \sqrt{|\Omega_1|^2 + |\Omega_2|^2 + |\Omega_3|^2}$. This gives the gauge potential [16]

$$\mathbf{A}_{11} = \hbar\left(\cos^2\phi\nabla S_{23} + \sin^2\phi\nabla S_{13}\right),$$
$$\mathbf{A}_{12} = \hbar\cos\theta\left(\frac{1}{2}\sin(2\phi)\nabla S_{12} - i\nabla\phi\right),$$
$$\mathbf{A}_{22} = \hbar\cos^2\theta(\cos^2\phi\nabla S_{13} + \sin^2\phi\nabla s_{23} \tag{5.45}$$

and the scalar potential

$$\mathscr{V}_{11} = \frac{\hbar^2}{2M}\sin^2\theta\left(\frac{1}{4}\sin^2(2\phi)(\nabla S_{12})^2 + (\nabla\phi)^2\right),$$
$$\mathscr{V}_{12} = \frac{\hbar^2}{2M}\sin\theta\left(\frac{1}{2}\sin(2\phi)\nabla S_{12} - i\nabla\phi\right)$$
$$\times\left(\frac{1}{2}\sin(2\phi)(\cos^2\phi\nabla S_{13} + \sin^2\phi\nabla S_{23}) - i\nabla\theta\right),$$
$$\mathscr{V}_{22} = \frac{\hbar^2}{2M}\left(\frac{1}{4}\sin^2(2\theta)(\cos^2\phi\nabla S_{13} + \sin^2\phi\nabla s_{23})^2 + (\nabla\theta)^2\right). \tag{5.46}$$

It is important to note at this stage that if the atoms are subject to any external trapping potential $W(\mathbf{r})$, then we also have to transform this potential to the dressed state basis. For the tripod system we therefore need to transform the potential $\hat{W}(\mathbf{r}) = W_1(\mathbf{r})|g_1\rangle\langle g_1| + W_2(\mathbf{r})|g_2\rangle\langle g_2| + W_3(\mathbf{r})|g_3\rangle\langle g_3|$ such that the 2×2 matrix has the matrix elements $W_{ik} = \langle D_i|\hat{W}(\mathbf{r})|D_k\rangle$. Using the expressions for the dark states, we obtain the effective scalar potential

$$W_{11} = W_2\cos^2\phi + W_1\sin^2\phi,$$
$$W_{12} = \frac{1}{2}(W_1 - W_2)\cos\theta\sin(2\phi),$$
$$W_{22} = (W_1\cos^2\phi + W_2\sin^2\phi)\cos^2\theta + W_3\sin^2\theta. \tag{5.47}$$

If each state in the tripod configuration feels the same trapping potential with $W_1(\mathbf{r}) = W_2(\mathbf{r}) = W_3(\mathbf{r})$, we see that the scalar potential is unchanged in the dressed state picture.

From (5.45), we see that by choosing the laser parameters, such as the phase and intensity ratios of the three laser beams, we can control the precise form of the gauge potential. Interestingly, it is possible to obtain non-trivial gauge potentials by only choosing plane waves with constant intensity. For instance, if the three Rabi frequencies have the same amplitude, with wave vectors

$$\mathbf{k}_1 = \frac{\kappa}{\sqrt{2}}\left(\mathbf{e}_x + \sqrt{3}\mathbf{e}_y\right),$$

$$\mathbf{k}_2 = \frac{\kappa}{\sqrt{2}}\left(\mathbf{e}_x - \sqrt{3}\mathbf{e}_y\right),$$

$$\mathbf{k}_3 = \sqrt{2}\kappa\mathbf{e}_x \tag{5.48}$$

with orthogonal polarisation and propagating in the xy plane at $120°$ from each other, then one obtains a gauge potential of the form [17]

$$\mathbf{A} = \frac{\hbar\kappa}{\sqrt{2}}(\sigma_x\mathbf{e}_x + \sigma_y\mathbf{e}_y). \tag{5.49}$$

This is an example of spin–orbit coupling of the Rashba–Dresselhaus type, where the σ_x and σ_y are the standard Pauli spin matrices, $\mathbf{e}_x, \mathbf{e}_y$ are orthonormal unit vectors given by the direction of the three wave vectors $\mathbf{k}_j$, and $\sqrt{2}\kappa = |\mathbf{k}_1| = |\mathbf{k}_2| = |\mathbf{k}_3|$. Other forms of gauge potentials can also be obtained; for instance, monopole fields, by choosing a combination of Laguerre–Gaussian laser beams with non-zero orbital angular momentum [16].

5.4.1   Spectrum

In order to understand better the properties of atoms that are subject to matrix gauge potentials, and how CIs can appear in the continuum regime, we need to look at the single particle spectrum of the system. We choose the symmetric gauge potential in (5.49) and solve the corresponding eigenvalue problem with the Schrödinger equation

$$E\bar{\Phi}(\mathbf{r}, t) = \left[\frac{(i\hbar\nabla - \mathbf{A}(\mathbf{r}))^2}{2M} + W(\mathbf{r}) + \mathcal{V}(\mathbf{r})\right]\bar{\Phi}(\mathbf{r}, t), \tag{5.50}$$

where $\bar{\Phi} = (\phi_1, \phi_2)^T$ is a two-component state.

For a homogeneous gas of atoms with a scalar potential $W(\mathbf{r})$ and trapping potential chosen such that $W(\mathbf{r}) + \mathcal{V}(\mathbf{r}) = 0$, the eigenvalues are readily calculated by Fourier transforming the eigenvalue equation, which results in the two bands

$$E_{\pm} = \frac{\hbar^2}{2M} \left[\left(k \pm \frac{\kappa}{\sqrt{2}} \right)^2 + \frac{3}{2}\kappa^2 \right], \tag{5.51}$$

where $k = |\mathbf{k}| = \sqrt{k_x^2 + k_y^2}$. The precise form of the spectrum depends on the details of the gauge potential. For the two-dimensional symmetric gauge potential chosen here, we see that for $|\mathbf{k}| = \kappa$ there is an infinite degeneracy, which can strongly influence the properties of Bose–Einstein condensation. For low momenta $|\mathbf{k}| \ll \kappa$, we can approximate the energy in (5.51) with

$$E_{\pm} = \pm \frac{\hbar^2}{2M} (\kappa k + 2\kappa^2), \tag{5.52}$$

which represents a Dirac cone for the atoms, see Fig. 5.4. For low momenta, we therefore also have access to quasi-relativistic dynamics. With a dispersion relation that is linear in momentum $\mathbf{k}$, the corresponding differential equation is the Dirac equation

$$i\hbar \frac{\partial}{\partial t} \Psi(\mathbf{r}, t) = \hbar \mathbf{A} \cdot \nabla \Psi(\mathbf{r}, t) \tag{5.53}$$

for a massless particle with the effective speed of light given by the coupling strength between the gauge potential and the atom. Interestingly, the effective speed of light can be several orders of magnitude smaller than the physical speed of light. This

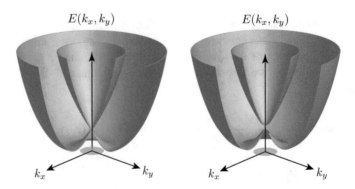

Fig. 5.4 The spectrum of spin–orbit coupled atoms (see also Fig. 4.5), where for small momenta, as indicated by the grey area, the dynamics is governed by a Dirac equation. A gap is opened up, illustrated in the figure to the right, if a scalar potential proportional to the σ_z Pauli spin matrix is present. This emulates a massive quasi-relativistic particle for small momenta

opens up the possibility to study relativistic phenomena in a condensed matter setting, but at ultralow temperatures and vastly different length- and time-scales compared to real relativistic situations.

5.4.2 A Quasi-Relativistic Example: The Atomic Zitterbewegung

Already in the early days of quantum mechanics the dynamics of relativistic particles was of key concern. It was soon realised that a number of counter-intuitive results would follow in the relativistic limit and one example is Zitterbewegung. This phenomenon is about a trembling motion of the centre of mass of a relativistic wave packet. The notion of Zitterbewegung has its roots in the work of Schrödinger, who studied the motion of a free particle based on Dirac's relativistic generalisation of a wave equation for spin-$\frac{1}{2}$ particles.

Zitterbewegung stems from the interference of positive and negative energy solutions of the Dirac equation for a free particle. The frequency of this interference process is determined by the energy gap between two possible solutions for the system. This energy gap is of the order of twice the rest energy $2m_ec^2$ of the electron, which is also the energy needed to create an electron–hole pair. If the Zitterbewegung relies on a process that corresponds to an energy of the order of the rest mass, then this unfortunately also makes it challenging to observe the trembling motion with real electrons. For synthetic matter, which obey a Dirac equation, this problem can be avoided. In the following, we assume the motion of the atoms to be restricted to one dimension, taken to be along the x-axis. A gas of ultracold atoms can be considered dynamically one-dimensional if the corresponding transversal energy scale given by the transversal trapping frequency is much higher than all other energy scales, such as the temperature or chemical potential in the presence of collisional interactions.

In this section we follow closely Ref. [18] where also a more in-depth and technical discussion can be found. To the effectively one-dimensional cloud of cold atoms, we apply the scheme for inducing non-Abelian gauge potentials. We are interested in the limit of low momenta, where the dynamics is effectively quasi-relativistic and described by a Dirac equation. For this purpose, we consider adiabatic motion of atoms in the presence of three laser beams. We choose a laser configuration where two of the beams have the same intensity but counter-propagate. The third laser beam has a different intensity compared to the two other laser beams. Its wave vector is chosen to be perpendicular to the axis defined by the propagation direction of laser 1 or 2. The laser configuration is shown in Fig. 5.5.

By defining the total Rabi frequency $\Omega = \sqrt{\sum_{n=1}^{3} |\Omega_n|^2}$ and the mixing angle $\tan\theta = \sqrt{|\Omega_1|^2 + |\Omega_2|^2}/|\Omega_3|$, we can write the Rabi frequencies as

$$\Omega_1 = \frac{\Omega}{\sqrt{2}} \sin\theta e^{-i\kappa x},$$

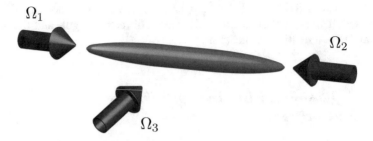

Fig. 5.5 One possible laser configuration for the tripod system which results in a non-trivial gauge potential for the two corresponding dark states

$$\Omega_2 = \frac{\Omega}{\sqrt{2}} \sin\theta e^{i\kappa x},$$

$$\Omega_3 = \Omega \cos\theta e^{-i\kappa y}. \tag{5.54}$$

With this choice of laser fields we obtain two dark states $|D_i\rangle$, $i = 1, 2$, given by

$$|D_1\rangle = \frac{1}{\sqrt{2}} e^{-i\kappa y} \left(e^{i\kappa x}|1\rangle - e^{-i\kappa x}|2\rangle \right),$$

$$|D_2\rangle = \frac{1}{\sqrt{2}} e^{-i\kappa y} \cos\theta \left(e^{ikx}|1\rangle - e^{-ikx}|2\rangle \right) - \sin\theta|3\rangle. \tag{5.55}$$

The corresponding two-component Schrödinger equation becomes

$$i\hbar \frac{\partial}{\partial t} \bar{\Psi} = \left[\frac{1}{2m} (\mathbf{p}_x - \mathbf{A})^2 + W + \mathscr{V} \right] \bar{\Psi} \tag{5.56}$$

with

$$\bar{\Psi} = \begin{pmatrix} \Psi_1 \\ \Psi_2 \end{pmatrix}. \tag{5.57}$$

Here, $\mathbf{p}_x = -i\hbar \mathbf{e}_x \partial_x$ denotes the momentum along the x-axis and m is the atomic mass. The scalar potential is defined by $W_{nm} = \langle D_n(\mathbf{r})|\hat{W}|D_m(\mathbf{r})\rangle$ with $\hat{W} = \sum_{j=1}^{3} W_j(\mathbf{r})|j\rangle\langle j|$ and $W_j(\mathbf{r})$ the trapping potential for atoms in the bare state j.

With the setup shown in Fig. 5.5 the gauge potential and scalar potentials in x-direction become

$$\mathbf{A} = -\hbar\kappa \begin{pmatrix} 0 & \mathbf{e}_x \cos\theta \\ \mathbf{e}_x \cos\theta & 0 \end{pmatrix} = -\hbar\kappa' \mathbf{e}_x \sigma_x,$$

$$\mathscr{V} = \frac{\hbar^2 \kappa^2}{2m} \begin{pmatrix} \sin^2\theta & 0 \\ 0 & \sin^2(2\theta)/4 \end{pmatrix},$$

$$W = \begin{pmatrix} W_1 & 0 \\ 0 & W_1 \cos^2\theta + W_3 \sin^2\theta \end{pmatrix}, \tag{5.58}$$

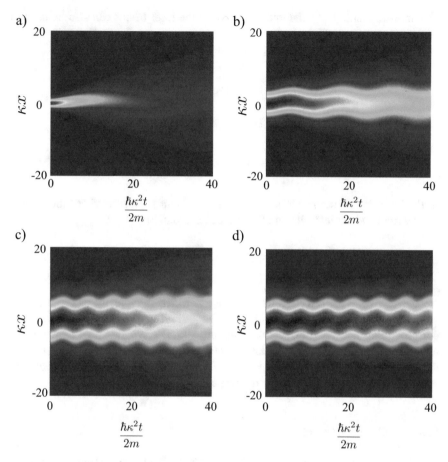

Fig. 5.6 Density plot showing the Dirac limit and the role of the initial width of the wave packets. (**a**)–(**c**) show the full Schrödinger dynamics using (5.56) for Gaussian initial states with increasing width σ. This results in a sharper momentum distribution with $p \ll \hbar\kappa$ increasingly fulfilled. The pure Dirac case is shown in (**d**) for comparison. The dynamics in (**a**)–(**d**) shows in addition to the Zitterbewegung also a damping

where we have introduced the notation $\kappa' = \kappa \cos(\theta)$ and assumed that the external trapping potentials for the first two atomic states are the same, i.e. $W_1 = W_2$.

5.4.2.1 The Dirac Limit

The existence of an energy gap is necessary in order to observe Zitterbewegung, see Fig. 5.6. Such a gap can be created by the different but constant potentials W_1 and W_3, which can be altered by detuning the corresponding lasers from the atomic transitions. Alternatively, the intensity ratio of the laser beams can also be used to adjust the scalar potential.

It is convenient to introduce the following notation:

$$V_z = \frac{1}{2}\Big[W_{11} + \mathcal{V}_{11} - (W_{22} + \mathcal{V}_{22})\Big] \tag{5.59}$$

and shift the zero level of energy. The trapping potential then reads

$$W + \mathcal{V} = V_z \sigma_z = V_z \begin{pmatrix} 1 & 0 \\ 0 & -1 \end{pmatrix}. \tag{5.60}$$

In the limit of low momenta, i.e. $|p| \ll \hbar\kappa$, we can neglect in (5.56) the kinetic energy term and are left with an effective Dirac equation

$$i\hbar\partial_t \bar{\Psi} = \left[-\frac{\mathbf{A}\cdot\mathbf{p}_x}{m} + \frac{\mathbf{A}^2}{2m} + V_z\sigma_z\right]\bar{\Psi} = \left[\tilde{c}\sigma_x\mathbf{p}_x + V_z\sigma_z\right]\bar{\Psi}. \tag{5.61}$$

Here, $\tilde{c} = \frac{\hbar\kappa}{m}$ is an effective recoil velocity, which is typically of the order of cm/s for alkali atoms. We can justify the Dirac limit in (5.61) by considering the corresponding linearised dispersion relation in (5.51). We also note that the $\mathbf{A}^2$-term can be absorbed into the scalar potential, so that we are left with an equation that resembles the Dirac equation for a free relativistic particle with the rest energy substituted by the potential energy difference between the two levels.

5.4.2.2   Zitterbewegung

Usually, Zitterbewegung is derived by solving the Heisenberg equation for the space operator. For a free particle with rest mass m, the Dirac Hamiltonian

$$\hat{H}_D = c\alpha\hat{\mathbf{p}} + \beta mc^2, \tag{5.62}$$

where c is the speed of light, is used to obtain the time-dependence for the position operator $\hat{\mathbf{x}}$. By using the anticommutation properties of the Dirac α and β matrices one finds

$$\hat{\mathbf{x}}(t) = \hat{\mathbf{x}}(0) + H_D^{-1}c^2\hat{\mathbf{p}}t - \frac{i\hbar c}{2}H_D^{-1}\left(e^{-2iH_Dt/\hbar} - 1\right)\left(\alpha(0) - c\hat{\mathbf{p}}H_D^{-1}\right). \tag{5.63}$$

Here, the first and second terms describe a motion that is linear in time, while the third term gives an oscillating contribution, which is the Zitterbewegung. To observe this trembling motion an initial 4-component spinor state needs to contain positive and negative parts as the α matrix is mixing these. The frequency of the oscillating term can be estimated in the particle's rest frame as $2mc^2/\hbar$. This typically large energy is the energy difference between a particle and its antiparticle (Fig. 5.7).

For cold atoms the two dark states that define our spin state are not a particle–antiparticle pair, but they are still separated by an energy gap, which is generated by the constant potential term in (5.60). In addition, the Dirac Hamiltonian now contains an effective rest mass V_z.

5.4.2.3 Dark State Dynamics

Our system with two degenerate dark states shows not only a relativistic behaviour, but also properties familiar from two-level systems in quantum optics. In order to see this in more detail, we write the spinor $\bar{\Psi}(x)$ as a combination of slowly varying envelopes $\phi_i(x)$, $i = 1, 2$, and coefficients that describe the population of the two dark states,

$$\bar{\Psi} = \begin{pmatrix} \phi_1(x)c_1(t) \\ \phi_2(x)c_2(t) \end{pmatrix}. \tag{5.64}$$

The spatial shape $\phi_i(x)$ should change much slower in time than the population $c_i(t)$ of the ith component of the spinor. The solutions are normalised according to $\langle \phi_i | \phi_i \rangle = 1$ and $|c_1|^2 + |c_2|^2 = 1$. After inserting the ansatz in (5.64) into Eq. (5.61), we obtain a set of coupled differential equations for the coefficients:

$$i\hbar \begin{pmatrix} \dot{c}_1 \\ \dot{c}_2 \end{pmatrix} = \begin{pmatrix} V_{z1} & \tilde{\Omega} \\ \tilde{\Omega}^* & V_{z2} \end{pmatrix} \begin{pmatrix} c_1 \\ c_2 \end{pmatrix}, \tag{5.65}$$

where

$$\tilde{\Omega} = \frac{\tilde{c}}{\hbar} \langle \phi_2 | \hat{\mathbf{p}}_x | \phi_1 \rangle, \tag{5.66}$$

and

$$V_{zi} = \frac{1}{\hbar} \langle \phi_i | V_z \hat{\sigma}_z | \phi_i \rangle. \tag{5.67}$$

The two spin components are coupled, hence the solutions of (5.65) show oscillations between the two dark states with a frequency

$$\omega_R^2 = |\tilde{\Omega}|^2 + \frac{1}{4} \left(V_{z1} - V_{z2} \right)^2. \tag{5.68}$$

For a vanishing $\tilde{\Omega}$, the coupling between the spin components in (5.65) becomes zero and we expect no oscillations of the populations. If the initial state has a non-

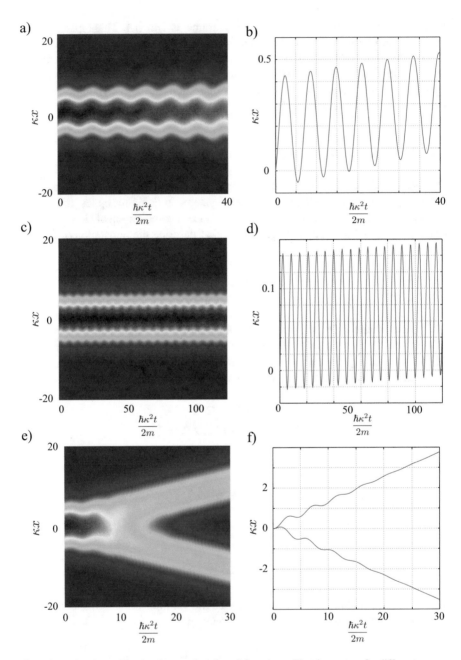

Fig. 5.7 Left column: The density as a function of time shows Zitterbewegung for different energy gaps ((**a**) with $V_z = \frac{\hbar^2 \kappa^2}{2m}$ and (**c**) with $V_z = 3\frac{\hbar^2 \kappa^2}{2m}$) and in (**e**) with a non-zero initial momentum $k_0 = \kappa'$. Right column: The centre of mass shows the expected oscillation ((**b**) and (**d**)) with an upward drift. With a non-zero initial momentum the behaviour is different, as can be seen in (**e**) and (**f**), where the Zitterbewegung breaks down after a few oscillations as the two states are moving in different directions. The initial spinor is in (**a**)–(**d**) $(1, 1)^T/\sqrt{2}$ and in (**e**) and (**f**) $(1, e^{i\pi/4})^T/\sqrt{2}$

zero momentum k_0, i.e.

$$\phi_i(x) = e^{-x^2/\sigma^2 + ik_0 x}, \tag{5.69}$$

where σ is the width, we obtain a non-zero $\tilde{\Omega}$ that consequently results in population transfer between the two dark states (Fig. 5.8).

5.4.2.4 Exact Solutions in the Schrödinger Limit

The atomic centre of mass motion is strictly speaking always governed by the Schrödinger equation (5.56). It is only in the limit of small momenta compared to the corresponding momentum contribution from the effective gauge potential, that we can describe the dynamics in terms of a Dirac equation. This would also require us to only use wave packets with zero momentum spread, which is unphysical. An exact solution for the full Schrödinger equation with a non-zero wave packet width is therefore needed in order to understand when the quasi-relativistic regime can be used [18].

The exact solution of the Schrödinger equation (5.56) with the gauge potentials from (5.58), can be written in momentum space,

$$\bar{\Psi}(k, \tau) = e^{-i(k^2 + 2\sigma_x k + \tilde{V}_z \sigma_z)\tau} \bar{\Psi}(k, 0), \tag{5.70}$$

where the dimensionless k is expressed in units of κ and the time τ in units of $2m/(\hbar \kappa^2)$. If we choose a Gaussian momentum distribution with a width Δ for the initial state,

$$\bar{\Psi}(k, 0) = \frac{1}{\sqrt{\Delta \sqrt{\pi}}} e^{-(k-k_0)^2/2\Delta^2} \begin{pmatrix} c_1 \\ c_2 \end{pmatrix} = \begin{pmatrix} \Psi_1(k, 0) \\ \Psi_2(k, 0) \end{pmatrix}, \tag{5.71}$$

the exact time-dependent solution is

$$\bar{\Psi}(k, \tau) = \frac{1}{\sqrt{\Delta \sqrt{\pi}}} e^{-\frac{(k-k_0)^2}{2\Delta^2} + i(k^2+1)\tau} \begin{pmatrix} c_1 \cos(\omega_k \tau) + \frac{i(c_1 \tilde{V}_z + c_2 2k)}{\omega_k} \sin(\omega_k \tau) \\ c_2 \cos(\omega_k \tau) - \frac{i(c_2 \tilde{V}_z - c_1 2k)}{\omega_k} \sin(\omega_k \tau) \end{pmatrix}, \tag{5.72}$$

where we have introduced the k-dependent frequency

$$\omega_k = \sqrt{4k^2 + \tilde{V}_z^2}. \tag{5.73}$$

With the solution $\bar{\Psi}(k, \tau)$, we can now calculate the centre of mass for the two-component wave packet, where we use the standard definition of the density $\rho(k, \tau) = |\Psi_1(k, \tau)|^2 + |\Psi_2(k, \tau)|^2$.

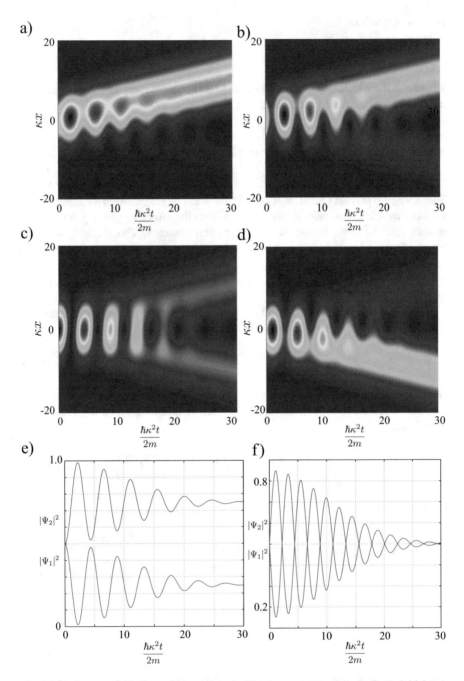

Fig. 5.8 Rabi type oscillations which emulate the Zitterbewegung can be seen for an initial state with non-zero momentum $k_0 = \kappa'$. The dynamics is sensitive to the initial conditions. (**a**) and (**b**) show $|\Psi_1(t,x)|^2$ for the initial spinor $(1,1)^T/\sqrt{2}$ and $(1, e^{i\pi/4})^T/\sqrt{2}$, respectively, whereas (**c**) and (**d**) show $|\Psi_2(t,x)|^2$ for the initial spinor $(1,1)^T/\sqrt{2}$ and $(1, e^{i\pi/4})^T/\sqrt{2}$. The population of the dark states is depicted in (**e**) with $(1,1)^T/\sqrt{2}$ and in (**f**) with $(1, e^{i\pi/4})^T/\sqrt{2}$

To illustrate this, we choose an initial state with $c_1 = c_2 = 1/\sqrt{2}$. The resulting populations of the two dark states are given by

$$|\psi_1(\tau)|^2 = \int_{-\infty}^{\infty} dk |\Psi_1(0, k)|^2 \left(1 + \frac{2k\tilde{V}_z}{\omega_k^2} \sin^2(\omega_k \tau) \right),$$

$$|\psi_2(\tau)|^2 = \int_{-\infty}^{\infty} dk |\Psi_2(0, k)|^2 \left(1 - \frac{2k\tilde{V}_z}{\omega_k^2} \sin^2(\omega_k \tau) \right). \tag{5.74}$$

The population difference, $\Delta N(t) = |\psi_1(t)|^2 - |\psi_2(t)|^2$, can easily be calculated in the limit $\Delta = 0$. In this limit, the Gaussian initial state is a delta function, and results in

$$\Delta N(\tau) = \lim_{\Delta \to 0} \int_{-\infty}^{\infty} dk \frac{4}{\Delta\sqrt{\pi}} e^{-\frac{(k-k_0)^2}{\Delta^2}} \frac{k\tilde{V}_z}{\omega_k^2} \sin^2(\omega_k \tau)$$

$$= \frac{4k_0\tilde{V}_z}{\sqrt{4k_0^2 + \tilde{V}_z^2}} \sin^2\left(\sqrt{4k_0^2 + \tilde{V}_z^2} \, \tau \right). \tag{5.75}$$

From this result, we see the importance of the initial momentum k_0. For $k_0 = 0$, there is no transfer of population between the dark states, whereas for a non-zero initial momentum the amplitude of the population oscillation is proportional to k_0. In addition, the frequency $\sqrt{4k_0^2 + \tilde{V}_z^2}$ also depends on k_0.

The expectation value for the centre of mass is calculated from the expression

$$\langle x(\tau) \rangle = i \int_{-\infty}^{\infty} dk \bar{\Psi}^\dagger(k, \tau) \partial_k \bar{\Psi}(k, \tau)$$

$$= \frac{1}{\Delta\sqrt{\pi}} \int_{-\infty}^{\infty} dk e^{-\frac{k^2}{\Delta^2}} \left(\frac{4k^2}{\omega_k^2} \tau + \frac{\tilde{V}_z^2}{\omega_k^3} \sin(2\omega_k \tau) \right). \tag{5.76}$$

From the first term under the integral sign, we obtain a drift term for the centre of mass,

$$x_d = \tau \left(1 - \sqrt{\pi} \frac{\tilde{V}_z}{\Delta} e^{\frac{\tilde{V}_z^2}{\Delta^2}} \text{Erfc}\left(\frac{\tilde{V}_z}{\Delta} \right) \right), \tag{5.77}$$

where Erfc is the complementary error function. In the limit of $\tilde{V}_z/\Delta \gg 1$, and using the asymptotic expansion of the error function

$$\text{Erfc}(x) = \frac{e^{-x^2}}{x\sqrt{\pi}} \sum_{n}^{\infty} (-1)^n \frac{(2n)!}{n!(2x)^{2n}}, \tag{5.78}$$

we obtain a reduced drift as a function of *increasing* $\tilde{V}_z/\Delta$. This is a finite size effect and is the result of having a finite width of the wave packet.

The second term under the integral in (5.76) represents the Zitterbewegung. In the limit $\tilde{V}_z/\Delta \gg 1$, the oscillating part in the integral gives

$$x_z = \frac{1}{\tilde{V}_z} \frac{\sin\left(2\tilde{V}_z\tau + \frac{1}{2}\arctan\left(\frac{\Delta^2}{4\tilde{V}_z}\tau\right)\right)}{\left(1 + \frac{\Delta^4}{16V_z^2}\tau^2\right)^{1/4}}. \tag{5.79}$$

From this expression, we see that a spread in the momentum distribution causes a damping also for the oscillating term of the centre of mass. The damping of the Zitterbewegung is relatively slow, but inevitable. The underlying equation is after all the Schrödinger equation, and only in the limit $\Delta = 0$ can we strictly speaking use a Dirac-type equation associated with no expansion of the wave packet, hence no damping. With the full Schrödinger equation a free wave packet always expands, albeit slowly if Δ is small, and hence also shows a damped Zitterbewegung.

For a typical alkali atom such as [87]Rb with a wave packet width of $10\,\mu$m, one would get $\Delta^2\tau/(4\tilde{V}_z) > 1$ for times larger than 1 ms, with a centre of mass oscillation frequency of the order of 1 kHz. Therefore, in order to detect the Zitterbewegung a broad initial wave packet is needed.

5.5   Cold Atoms and the Bose–Einstein Condensate

In 1995, experimentalists were able to cool and trap gases of [87]Rb [19, 20] and [7]Li [21] to such low temperatures that quantum effects started to play a major role. These gases consisted of bosonic atoms, which meant that they formed a Bose–Einstein condensate (BEC) with exotic properties such as superfluidity and many-body effects at ultracold temperatures. It is fair to say that nobody anticipated the impact of the arrival of this new quantum fluid. Today, the BEC is used as a tool to study exotic properties of matter by creating macroscopic quantum states and using these for probing effects such as quantum phase transitions and nonequilibrium features in quantum gases. One of the most appealing factors when using ultracold gases is the interplay between quantum optics and condensed matter physics. Also, these gases allow for an unprecedented control over the physical parameters in experiments, ranging from condensate density, trap geometry, and even collisional strength using Feshbach resonances. In practice, this also means that theory and experiments are in a very fruitful symbiosis. The available theoretical tools often describe the dynamics of the gas extremely well. This has opened up the possibility to address the concept of quantum simulators in these systems, originally proposed by Feynman [13], where the physical processes in an experiment is well described by some many-body Hamiltonian that may be too difficult to analyse on a classical computer due to the large Hilbert space. But the experiment, on the other hand, does not care about the Hilbert space. The experiment could allow, for instance, the quantum fluid to relax to its ground states, and, by doing so, provide us the numerically or analytically inaccessible ground state.

In this section, we review some standard theoretical tools for analysing weakly interacting BEC. The emphasis is on the mean-field treatment, and how we can obtain an effective equation of motion for the gas. This is the celebrated Gross–Pitaevskii equation.

5.5.1   The Description of a Condensate

A BEC can be described in terms of a macroscopically occupied single quantum state. Such a many-body quantum gas can in the non-interacting case be analysed using standard techniques from thermodynamics. However, in most situations we need to also take into account collisions between the atoms. This will affect the properties of the gas significantly. In the following we will consider a weakly interacting gas. It is therefore natural to rely on a mean-field treatment of the condensate. The inclusion of collisions between the atoms does in fact not change too much the *onset* of BEC and the corresponding critical temperature, compared to the non-interacting gas. At low temperatures, however, the collisions do affect the condensate as far as ground state properties and dynamics are concerned.

It is often a good approximation to assume zero temperature for the cloud of atoms. With state of the art cooling techniques it is possible to reach temperatures far below the critical temperature, which is typically in the μK regime, where any contribution from the remaining thermal component can be neglected, as one can see in Fig. 5.9. For a dilute gas, two-body collisions will be the dominant process. In addition, because the gas is cold, we need to consider only s-wave scattering as the mechanism for the interaction. The interaction potential is therefore of the form [23]

$$V_{\text{int}}(r - r') = \frac{4\pi \hbar^2 a}{m} \delta(r - r'), \tag{5.80}$$

where the interaction is described by the single parameter a called the s-wave scattering length. For a derivation of (5.80), we have to solve the two-body scattering problem in the limit of zero momentum.

With these assumptions, we obtain a Hamiltonian of the form

$$\hat{H} = \int d\mathbf{r} \left\{ \hat{\psi}^{\dagger}(\mathbf{r}) \left(-\frac{\hbar^2}{2m} \nabla^2 + V_{\text{ext}}(\mathbf{r}) \right) \hat{\psi}(\mathbf{r}) + \frac{g}{2} \hat{\psi}^{\dagger}(\mathbf{r}) \hat{\psi}^{\dagger}(\mathbf{r}) \hat{\psi}(\mathbf{r}) \hat{\psi}(\mathbf{r}) \right\},$$
$$\tag{5.81}$$

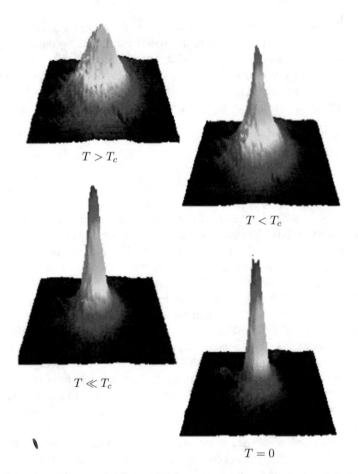

Fig. 5.9 The onset of BEC as a function of temperature T is seen as a sharp peak in the density in the centre of the confining trap. The right lower picture shows an almost pure condensate where only a small thermal component can be seen. Pictures, courtesy of A. Arnold, are from the BEC experiment at University of Strathclyde, Glasgow, UK [22]

where $g = 4\pi\hbar^2 a/m$ and V_{ext} is an external trapping potential. The field operators $\hat{\Psi}(\mathbf{r}, t)$ and $\hat{\Psi}^\dagger(\mathbf{r}, t)$ destroy and create, respectively, a particle at $\mathbf{r}$ at time t, and they obey the usual bosonic commutation rules

$$[\hat{\Psi}(\mathbf{r}), \hat{\Psi}^\dagger(\mathbf{r}')] = \delta(r - r')\delta(t - t') \qquad (5.82)$$

and

$$[\hat{\Psi}(\mathbf{r}, t), \hat{\Psi}(\mathbf{r}', t')] = [\hat{\Psi}^\dagger(\mathbf{r}, t), \hat{\Psi}^\dagger(\mathbf{r}', t')] = 0. \qquad (5.83)$$

With these commutation rules, we get the Heisenberg equation of motion for the field operator

$$i\hbar \frac{\partial}{\partial t}\hat{\Psi} = [\hat{H}, \hat{\Psi}] = -\frac{\hbar^2}{2m}\nabla^2\hat{\Psi} + V_{\text{ext}}(\mathbf{r})\hat{\Psi} + g\hat{\Psi}^{\dagger}\hat{\Psi}^{\dagger}\hat{\Psi}. \tag{5.84}$$

We split the field operator $\hat{\Psi}$ into the operator for the lowest mode and a part representing the fluctuations and thermal excitations,

$$\hat{\Psi}(\mathbf{r}) = \hat{\Psi}_0(\mathbf{r}) + \delta\hat{\Psi}(\mathbf{r}). \tag{5.85}$$

At zero temperature we can as a first approximation neglect the fluctuations $\delta\hat{\Psi}(\mathbf{r})$. In the presence of a condensate, we note that the lowest mode is macroscopically populated. We can therefore write

$$\hat{\Psi}(\mathbf{r}) = \Psi(\mathbf{r})\hat{a}_0 \approx \Psi(\mathbf{r})\sqrt{N}, \tag{5.86}$$

where we have replaced the annihilation operator $\hat{a}_0$ by $\sqrt{N}$, which is often called the Bogoliubov approximation[24]. This is a legitimate approximation provided that the number of atoms N in the condensate is sufficiently large. What we have done is nothing but replacing the field operator by its average

$$\hat{\Psi}(\mathbf{r}) \approx \langle\hat{\Psi}(\mathbf{r})\rangle = \Psi(\mathbf{r})\sqrt{N}. \tag{5.87}$$

The resulting equation of motion for the condensate wave function $\Psi(\mathbf{r})$ then becomes

$$i\hbar \frac{\partial}{\partial t}\Psi(\mathbf{r}, t) = \left(-\frac{\hbar^2}{2m}\nabla^2 + V_{\text{ext}}(\mathbf{r}) + g|\Psi(\mathbf{r}, t)|^2\right)\Psi(\mathbf{r}, t). \tag{5.88}$$

This is the Gross–Pitaevskii equation [24], which is the true workhorse for describing the dynamics of a Bose–Einstein condensate. The Gross–Pitaevskii equation is based on mean-field theory, in which each atom feels the presence of all the other atoms through an effective potential which is proportional to the density of the cloud. It describes the dynamics of a condensate remarkably well and has been used extensively in studies of BECs.

The time-independent version of Eq. (5.88),

$$\mu\varphi(\mathbf{r}) = \left(-\frac{\hbar^2}{2m}\nabla^2 + V_{\text{ext}}(\mathbf{r}) + g|\varphi(\mathbf{r})|^2\right)\varphi(\mathbf{r}), \tag{5.89}$$

is obtained using the Ansatz $\Psi(\mathbf{r}, t) = \varphi(\mathbf{r})e^{-i\mu t\hbar}$, where μ is the chemical potential. The external potential $V_{\text{ext}}(\mathbf{r})$ can take many shapes. It can stem from magnetic trapping, but also from the optical traps discussed in the previous section.

5.5.2 Conical Intersections and the Gross–Pitaevskii Equation

At the Gross–Pitaevskii mean-field level, the order parameter is given by a non-linear Schrödinger type equation. As we have seen, for atoms with internal hyperfine structure, the order parameter becomes a spinor, e.g. for two-level atoms

$$\Psi(\mathbf{x}, t) = \begin{bmatrix} \psi_+(\mathbf{x}, t) \\ \psi_-(\mathbf{x}, t) \end{bmatrix}. \tag{5.90}$$

Correspondingly, the Gross–Pitaevskii differential equation (5.88) becomes matrix-valued. For brevity, let us consider the non-linear version of the Rashba Hamiltonian (4.59) here, i.e.

$$i\hbar \frac{\partial}{\partial t} \begin{bmatrix} \psi_+(\mathbf{x}, t) \\ \psi_-(\mathbf{x}, t) \end{bmatrix} = \left\{ -\frac{\hbar^2}{2m} \left(\frac{\partial^2}{\partial x^2} + \frac{\partial^2}{\partial y^2} \right) \right.$$

$$\left. -i\hbar\alpha_R \left(\sigma_y \frac{\partial}{\partial x} - \sigma_x \frac{\partial}{\partial y} \right) + g|\psi_+(\mathbf{x}, t)|^2 + g|\psi_-(\mathbf{x}, t)|^2 \right\} \begin{bmatrix} \psi_+(\mathbf{x}, t) \\ \psi_-(\mathbf{x}, t) \end{bmatrix}. \tag{5.91}$$

We consider only *density–density* interactions between the hyperfine states of the atoms, i.e. the atom does not change its internal state upon scattering with another atom. This is motivated noticing that such processes are usually negligible for most atoms and relevant energy scales. Furthermore, for simplicity we also set the interaction strengths to be the same between the different levels.

The Gross–Pitaevskii equation may, in general, be written as

$$i\hbar \frac{\partial}{\partial t} \Psi(\mathbf{x}, t) = H_{\mathrm{GP}}\left(\mathbf{x}, |\Psi(\mathbf{x}, t)|^2 \right) \Psi(\mathbf{x}, t), \tag{5.92}$$

where we have defined the 'Hamiltonian' $H_{\mathrm{GP}}\left(\mathbf{x}, |\Psi(\mathbf{x}, t)|^2 \right)$, which depends on the state $\Psi(\mathbf{x}, t)$. Since the equation is non-linear, the superposition principle no longer holds, and we cannot in a strict sense talk about concepts like adiabaticity and Berry phase. Nevertheless, we may talk about stationary states defined by

$$H_{\mathrm{GP}}\left(\mathbf{x}, |\Psi_n(\mathbf{x})|^2 \right) \Psi_n(\mathbf{x}) = \mu_n \Psi_n(\mathbf{x}), \tag{5.93}$$

where n labels the state and its energy μ_n. In the limit of vanishing interaction, $g \to 0$, the stationary states become the eigenstates of the corresponding single particle Hamiltonian. Thus, for the Rashba Gross–Pitaevskii model (5.91), we must have that for $g = 0$ the energies $\mu_\pm(p) = E_\pm(p)$ with $E_\pm(p)$ given in (4.60), and $p = |\mathbf{p}|$. An intersection for $p_y = 0$ of these surfaces is shown in Fig. 5.10a as dotted lines, i.e. we reproduce the Dirac CI already plotted in Fig. 4.5a. As we have

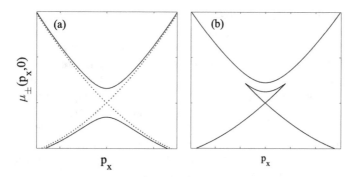

Fig. 5.10 Schematic picture of the energies $\mu_\pm(p)$ (5.93) for the Rashba Gross–Pitaevskii model (5.91). In both plots, we show the $p_y = 0$ intersection of the two-dimensional surfaces. In (**a**), the interaction is zero, i.e. $g = 0$, and thereby the non-linearity vanishes. The solid lines give the energies with a non-zero Zeeman field $\lambda\sigma_z$, while the dotted lines are the regular Dirac CI also seen in Fig. 4.5a. In (**b**), we consider the same model, but with interaction $g \neq 0$, and $\lambda \neq 0$. For sufficiently strong interaction $|g| > g_c > 0$ (g_c a critical interaction strength), a loop structure, in the shape of a 'swallowtail', is formed. The sign of the interaction strength (attractive vs. repulsive) determines whether the loop forms on the upper or lower branch

discussed, if we add a *Zeeman term* (or 'mass term') $\lambda\hat\sigma_z$ to the model, the Dirac CI splits up,

$$\mu_\pm(p) = \frac{p^2}{2m} \pm \sqrt{\lambda^2 + (\alpha_R p)^2}. \tag{5.94}$$

This is demonstrated as the solid lines in Fig. 5.10a. So far there is nothing new since the interaction is set to zero. What about non-zero interactions?

In Fig. 5.10b, we show the schematic energies (not numerically calculated) for non-zero interaction and a non-zero λ. Two new solutions appear when $|g| > g_c$ for some critical coupling strength $g_c > 0$. More precisely, a loop structure emerges in the spectrum. Due to its shape, these have been named *swallowtail loops* [25]. They were first studied in the Landau–Zener model [26], and later verified experimentally with cold atoms loaded in an optical lattice [27]. The formation of swallowtail loops also for Dirac CI was first considered for a honeycomb lattice [28] and more recently for the Rashba-type spin–orbit coupling [29]. The loop is a manifestation of hysteresis deriving from the non-linearity of the model. To see this, assume the system starts in a stationary state on the lower dispersion with a momentum $(p_x, p_y) = (-p_0, 0)$ and a weak linear potential $V(x, y) = -Fx$ is applied such that the $(p_x, p_y) = (-p_0 + Ft, 0)$. The state thus first approaches the CI from the left, passes through it and continues upwards to the right. However, at some point it reaches the tip of the swallowtail loop and can no longer follow the dispersion; it has to make a discontinuous 'jump,' where a part populates the upper and a part populates the lower branch. Now, if the force is reversed, the system is not able to return to its original state, i.e. it forms a hysteresis loop. Due to the non-linearity, one says that the system is no longer able to pass adiabatically through the CI, unless it resides on the upper branch [27].

After all, the Gross–Pitaevskii equation is an approximation to an interacting many-body problem, and the full model is of course linear, meaning that loops like the one seen in Fig. 5.10 are not possible. It turns out, though, that the many-body spectrum still resembles the structure of the mean-field dispersions [30], and the non-adiabatic behaviour would still be very similar to that explained above for the hysteresis [27]. In the many-body setting there are numerous avoided crossings of energy levels, but the energy gaps of most of these crossings scale as $1/N$ with N the particle number. Hence, in the limit $N \to \infty$ it becomes impossible for the system to stay adiabatic and follow a single adiabatic states.

References

1. Greiner, M., Mandel, O., Esslinger, T., Hänsch, T.W., Bloch, I.: Quantum phase transition from a superfluid to a Mott insulator in a gas of ultracold atoms. Nature **415**, 39 (2002)
2. Bartenstein, M., Altmeyer, A., Riedl, S. Jochim, S., Chin, C., Denschlag, J.H., Grimm, R.: Crossover from a molecular Bose-Einstein condensate to a degenerate Fermi gas. Phys. Rev. Lett. **92**, 120401 (2004)
3. Zwierlein, M.W., Stan, C.A., Schunck, C.H., Raupach, S.M.F., Kerman, A.J., Ketterle, W.: Condensation of pairs of fermionic atoms near a Feshbach resonance. Phys. Rev. Lett. **92**, 120403 (2004)
4. Kinast, J., Hemmer, S.L., Gehm, M.E., Turlapov, A., Thomas, J.E.: Evidence for superfluidity in a resonantly interacting Fermi gas. Phys. Rev. Lett. **92**, 150402 (2004)
5. Bourdel, T., Khaykovich, L., Cubizolles, J., Zhang, J., Chevy, F., Teichmann, M., Tarruell, L., Kokkelmans, S., Salomon, C.: Experimental study of the BEC-BCS crossover region in Lithium 6. Phys. Rev. Lett. **93**, 050401 (2004)
6. Bissbort, U., Götze, S., Li, Y., Heinze, J., Krauser, J.S., Weinberg, M., Becker, C., Sengstock, K., Hofstetter, W.: Detecting the amplitude mode of strongly interacting lattice Bosons by Bragg scattering. Phys. Rev. Lett. **106**, 205303 (2011)
7. Endres, M., Fukuhara, T., Pekker, D., Cheneau, M., Schauß, P., Gross, C., Demler, E., Kuhr, S., Bloch, I.: The 'Higgs' amplitude mode at the two-dimensional superfluid/Mott insulator transition. Nature **487**, 454 (2012)
8. Merkl, M., Jacob, A., Zimmer, F.E., Öhberg, P., Santos, L.: Chiral confinement in quasirelativistic Bose-Einstein condensates. Phys. Rev. Lett. **104**, 073603 (2010)
9. Bermudez, A., Mazza, L., Rizzi, M., Goldman, N., Lewenstein, M., Martin-Delgado, M.A.: Wilson Fermions and axion electrodynamics in optical lattices. Phys. Rev. Lett. **105**, 190404 (2010)
10. Williams, J.R., Hazlett, E.L., Huckans, J.H., Stites, R.W., Zhang, Y., O'Hara, K.M.: Evidence for an Excited-State Efimov trimer in a three-component Fermi gas. Phys. Rev. Lett. **103**, 130404 (2009)
11. Wilczek, F., Zee, A.: Appearance of gauge structure in simple dynamical systems. Phys. Rev. Lett. **52**, 2111 (1984)
12. Lewenstein, M., Sanpera, A., Ahufinger, V., Damski, B., Sen De, A., Sen, U.: Ultracold atomic gases in optical lattices: mimicking condensed matter physics and beyond. Adv. Phys. **56**, 243 (2007)
13. Feynman, R.P.: Simulating physics with computers. Int. J. Theor. Phys. **21**, 467 (1982)
14. Goldman, N., Juzeliūnas, G., Öhberg, P., Spielman, I.B.: Light-induced gauge fields for ultracold atoms. Rep. Prog. Phys. **77**, 126401 (2014)

15. Stenholm, S.: The semiclassical theory of laser cooling. Rev. Mod. Phys. **58**, 699 (1986)
16. Ruseckas, J., Juzeliūnas, G., Öhberg, P., Fleischhauer, M.: Non-Abelian gauge potentials for ultracold atoms with degenerate dark states. Phys. Rev. Lett. **95**, 010404 (2005)
17. Juzeliūnas, G., Ruseckas, J., Dalibard, J.: Generalized Rashba-Dresselhaus spin-orbit coupling for cold atoms. Phys. Rev. A **81**, 053403 (2010)
18. Merkl, M., Zimmer, F.E., Juzeliūnas, G., Öhberg, P.: Atomic Zitterbewegung. EPL **83**, 54002 (2008)
19. Anderson, M.H., Ensher, J.R., Matthews, M.R., Wieman, C.E., Cornell, E.A.: Observation of Bose-Einstein condensation in a dilute atomic vapor. Science **269**, 198 (1995)
20. Davis, K.B., Mewes, M.-O., Andrews, M.R., van Druten, N.j., Durfee, D.S., Kurn, D.M., Ketterle, W.: Bose-Einstein condensation in a gas of sodium atoms. Phys. Rev. Lett. **75**, 3969 (1995)
21. Bradley, C.C., Sackett, C.A., Tollett, J.J., Hulet, R.G.: Evidence of Bose-Einstein condensation in an atomic gas with attractive interactions. Phys. Rev. Lett. **75**, 1687 (1995)
22. Arnold, A.S., Garvie, C.S., Riis, E.: Large magnetic storage ring for Bose-Einstein condensates. Phys. Rev. A **73**, 041606 (2006)
23. Huang, K.: Statistical Mechanics. Wiley, New York (1987)
24. Pitaevskii, L.-P., Stringari, S.: Bose-Einstein Condensation. Clarendon Press, Oxford (2003)
25. Machholm, M., Pethick, C.J., Smith, H.: Band structure, elementary excitations, and stability of a Bose-Einstein condensate in a periodic potential. Phys. Rev. A **67**, 053613 (2003)
26. Wu, B., Niu, Q.: Nonlinear Landau-Zener tunneling. Phys. Rev. A **61**, 023402 (2000)
27. Chen, Y.-A., Huber, S.D., Trotzky, S., Bloch, I., Altman, E.: Many-body Landau-Zener dynamics in coupled one-dimensional Bose liquids. Nature Phys. **7**, 61 (2011)
28. Chen, Z., Wu, B.: Bose-Einstein condensate in a honeycomb optical lattice: fingerprint of superfluidity at the Dirac point. Phys. Rev. Lett. **107**, 065301 (2011)
29. Zhang, Y., Gui, Z., Chen, Y.: Nonlinear dynamics of a spin-orbit-coupled Bose-Einstein condensate. Phys. Rev. A **99**, 023616 (2019)
30. Witthaut, D., Graefe, E.M., Korsch, H.J.: Towards a generalized Landau-Zener formula for an interacting Bose-Einstein condensate in a two-level system. Phys. Rev. A **73**, 063609 (2006)

Chapter 6
Conical Intersections in Other Physical Systems

Abstract In this chapter, we collect a few other physical scenarios where conical intersections (CIs) appear. We demonstrate the existence of CIs in an atom-cavity system, by treating the quadrature variables as slow parameters. Similar features can be found in ion trap systems, where the cavity field is replaced by the phonons associated with the trapping potential. Common to these systems are the quantum Rabi and Jaynes–Cummins models that naturally describe the interaction between the discrete energy levels of the atom or ion with the oscillator fields associated with the photons in the cavity or the phonons of the trap. We discuss how CIs may show up for classical light moving through materials with a small space-dependent refractive index and how lattice models can be emulated in such settings. These systems may realise Dirac cones by appropriately choosing the lattice structure. We finally examine the exceptional point (EP) concept for parameter-dependent non-Hermitian systems, which exhibit complex-valued energy spectra. An EP is a crossing point in parameter space between two or more energies; thus, an EP is the non-Hermitian counterpart to a CI for Hermitian system. We discuss how non-Hermitian systems can appear as effective models of open quantum systems.

6.1 Cavity Quantum Electrodynamics

In Sects. 3.3 and 3.4, we discussed a family of models hosting CIs between their APSs. A common feature among these JT models is that they describe harmonic oscillators coupled to some two-level system. For molecules, the harmonicity is an approximation and anharmonic corrections are unavoidable. As we have seen, these corrections are typically not altering the topological property of the models, e.g. the Berry phase assigned to a two-dimensional CI remains to be $\pm\pi$. As we will see next, and in the following section, harmonic oscillators coupled to few-level systems naturally occur in other settings, and equivalent physical features emerge.

The electromagnetic field when quantised is described by a set of harmonic oscillators, each corresponding to one particular photon mode with a wavelength λ expressed in the oscillator frequency ω as $\lambda = 2\pi c/\omega$, where c is the speed of light. We will not go through the actual quantisation of the electromagnetic

© Springer Nature Switzerland AG 2020

J. Larson et al., *Conical Intersections in Physics*, Lecture Notes in Physics 965,
https://doi.org/10.1007/978-3-030-34882-3_6

radiation field here, but instead accept it as such and refer the interested reader to any book on quantum optics, for instance [1]. In free space, the frequencies form a continuum, while inside a perfectly conducting resonator only certain modes are allowed as given by the associated boundary conditions. For instance, a Fabry–Pérot cavity consists of two almost perfectly reflecting parallel mirrors that form the resonator. The longitudinal modes between the mirrors are to a good approximation standing waves with well-defined wave numbers. The idea of cavity quantum electrodynamics (QED) is to use a high-Q cavity with very long photon lifetime, isolate one or a few of the photon modes, and couple them coherently to matter.

6.1.1   The Jaynes–Cummings and Quantum Rabi Models

Within the dipole approximation and when we consider a single atom confined in the cavity, the Hamiltonian may be decomposed as

$$\hat{H} = \hat{H}_{\mathrm{f}} + \hat{H}_{\mathrm{a}} - e\hat{\mathbf{r}} \cdot \hat{\mathbf{E}}. \tag{6.1}$$

The first term is the Hamiltonian of the free electromagnetic field, for which we only include a single mode, i.e.

$$\hat{H}_{\mathrm{f}} = \hbar\omega\hat{a}^{\dagger}\hat{a}, \tag{6.2}$$

where we have neglected the zero-point energy term and $\hat{a}$ ($\hat{a}^{\dagger}$) is the photon annihilation (creation) operator obeying the boson commutation relation $[\hat{a}, \hat{a}^{\dagger}] = 1$. With $|n\rangle$ an n-photon Fock state, we have $\hat{a}|n\rangle = \sqrt{n}|n - 1\rangle$, $\hat{a}^{\dagger}|n\rangle = \sqrt{n + 1}|n + 1\rangle$, and $\hat{n}|n\rangle = n|n\rangle$, where $\hat{n} = \hat{a}^{\dagger}\hat{a}$ is the photon number operator. The second term of (6.1) is the free atomic Hamiltonian, and we assume that we can restrict it to only two electronic hyperfine levels $|\pm\rangle$ (*two-level approximation*), i.e. all other transitions are either forbidden by selection rules or far off resonant. We also disregard the motion of the atom. If $\hbar\Omega$ is the energy difference between the two electronic states, we can write

$$\hat{H}_{\mathrm{a}} = \frac{\hbar\Omega}{2}\hat{\sigma}_z \tag{6.3}$$

with $\hat{\sigma}_z|\pm\rangle = \pm|\pm\rangle$. The last term in the Hamiltonian (6.1) represents the dipole coupling between the atom and the radiation field. Within the two-level approximation, we have $\hat{\mathbf{r}} \sim \hat{\sigma}_x$, where $\hat{\sigma}_x|\pm\rangle = |\mp\rangle$. The electric field in the single-mode and dipole approximations takes the form $\hat{\mathbf{E}} \propto (\hat{a} + \hat{a}^{\dagger})$. Adding all this together we arrive at the *quantum Rabi model*

$$\hat{H}_{\mathrm{qRabi}} = \hbar\omega\hat{a}^{\dagger}\hat{a} + \frac{\hbar\Omega}{2}\hat{\sigma}_z + \hbar g\hat{\sigma}_x\left(\hat{a} + \hat{a}^{\dagger}\right), \tag{6.4}$$

where the atom–light coupling is given by $g = -\langle -|\hat{\mathbf{r}}|+\rangle \cdot \epsilon \mathscr{E}/\hbar$ with the field amplitude $\mathscr{E} = \sqrt{\hbar\omega/2\epsilon_0 V}$, ϵ the polarisation vector, and V the effective mode volume.

We have imposed a series of approximations to derive the quantum Rabi Hamiltonian (6.4); single-mode, two-level, dipole, and neglect of atomic motion. These are often justified in experiments, as is the *rotating wave approximation* (RWA), which omits virtual terms in the Hamiltonian. To see this, we first express the Pauli-x operator in raising ($\hat{\sigma}^+$) and lowering ($\hat{\sigma}^-$) atomic operators, i.e. $\hat{\sigma}_x = \hat{\sigma}^+ + \hat{\sigma}^-$, in terms of which the interaction term can be written as

$$g\left(\hat{\sigma}^+ + \hat{\sigma}^-\right)\left(\hat{a} + \hat{a}^\dagger\right) = g\left(\hat{\sigma}^+\hat{a} + \hat{a}^\dagger\hat{\sigma}^- + \hat{\sigma}^+\hat{a}^\dagger + \hat{\sigma}^-\hat{a}\right). \tag{6.5}$$

The meaning of the four terms should be clear: $\hat{\sigma}^+\hat{a}$—the atom's electronic state is excited by absorbing a photon; $\hat{a}^\dagger\hat{\sigma}^-$—the reverse process where the atom is de-excited by emitting a photon; $\hat{\sigma}^+\hat{a}^\dagger$—a virtual process where the atom gets excited by emitting a photon, and finally another virtual process $\hat{\sigma}^-\hat{a}$—the atom de-excites by absorbing a photon. The first two of these are schematically pictured in Fig. 6.1. If we turn to an interaction picture with respect to $\hat{H}_f + \hat{H}_a$, we have that the first two terms acquire phase factors $e^{\pm i(\omega-\Omega)t}$ and the last two terms acquire phase factors $e^{\pm i(\omega+\Omega)t}$. In any cavity QED experiment $|g| \ll \omega$, and we may safely neglect the *counter-rotating terms* $\hat{\sigma}^+\hat{a}^\dagger + \hat{\sigma}^-\hat{a}$. By doing so, we note that the number of excitations, as represented by the operator $\hat{N} = \hat{n} + \frac{1}{2}\hat{\sigma}_z$, is preserved. By neglecting the virtual terms we find the *Jaynes–Cummings Hamiltonian*

$$\hat{H}_{JC} = \hbar\omega\hat{n} + \frac{\hbar\Omega}{2}\hat{\sigma}_z + \hbar g\left(\hat{\sigma}^+\hat{a} + \hat{a}^\dagger\hat{\sigma}^-\right). \tag{6.6}$$

It turns out practical to switch from the annihilation–creation operator representation to the *quadrature representation* for the field mode, meaning that we express them in the conjugate variables as

$$\hat{x} = \frac{1}{\sqrt{2}}\left(\hat{a}^\dagger + \hat{a}\right), \qquad \hat{p} = \frac{i}{\sqrt{2}}\left(\hat{a}^\dagger - \hat{a}\right). \tag{6.7}$$

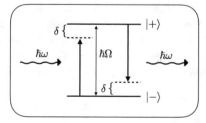

Fig. 6.1 The light–matter interaction processes of the Jaynes–Cummings model. The internal atomic state is excited or de-excited, $|\mp\rangle \to |\pm\rangle$, by absorbing or emitting a photon of energy $\hbar\omega$. The energy difference between the two atomic states is $\hbar\Omega$, while the detuning is $\delta = \hbar\Omega - \hbar\omega$

The Jaynes–Cummings Hamiltonian in the quadrature representation takes the form

$$\hat{H}_{\text{JC}} = \hbar\omega \left(\frac{\hat{p}^2}{2} + \frac{\hat{x}^2}{2} \right) + \frac{\hbar\Omega}{2}\hat{\sigma}_z + \frac{\hbar g}{\sqrt{2}} \left(\hat{x}\hat{\sigma}_x - \hat{p}\hat{\sigma}_y \right), \tag{6.8}$$

where $\hat{\sigma}_y$ is the Pauli y-matrix with $\hat{\sigma}_y|\pm\rangle = \pm i|\pm\rangle$. In the language of 'position' $\hat{x}$ and 'momentum' $\hat{p}$, the Jaynes–Cummings interaction is nothing that we have seen so far—it couples the 'spin' to both the position (like in molecular models of Chap. 3) and the momentum (like in models from condensed matter physics discussed in Chap. 4, and in cold atom physics, discussed in Chap. 5). We may impose a BOA by diagonalising the spin degrees of freedom,

$$\hat{H}_{\text{JC}}^{(ad)} = \hbar\omega \left(\frac{\hat{p}^2}{2} + \frac{\hat{x}^2}{2} \right) + \hbar\sqrt{\frac{\Omega^2}{4} + \frac{g^2}{2} \left(\hat{x}^2 + \hat{p}^2 \right)}\hat{\tau}_z, \tag{6.9}$$

with $\hat{\tau}_z$ is the corresponding Pauli z-matrix in the adiabatic basis. In the crudest mean-field approach we replace the quadrature operators by c-numbers. We then find the semi-classical adiabatic energy

$$E_{\text{scJC}}^{(\pm)}(x, p) = \hbar\omega \left(\frac{p^2}{2} + \frac{x^2}{2} \right) \pm \hbar\sqrt{\frac{\Omega^2}{4} + \frac{g^2}{2} \left(x^2 + p^2 \right)}. \tag{6.10}$$

We see that these two energy surfaces in the xp-plane reproduce the by now well-known structure like in Fig. 4.5b. However, the non-zero Ω split the CI and opens up an energy gap. Note that for $\Omega \geq 2g^2/\omega$, the lower energy surface $E_{\text{scJC}}^{(-)}(x, p)$ loses the sombrero shape and a global minimum at $(x, p) = (0, 0)$ occurs. While this view relies on a semi-classical mean-field approach, there is still a non-vanishing Berry phase that can be associated with this CI [2].

6.1.2   The Intrinsic Anomalous Hall Effect in Cavity QED

In modern setups of *circuit QED*, the light–matter coupling strength g can become close to the photon frequency ω [3] and the RWA breaks down. By including the counter-rotating terms to regain the quantum Rabi model (6.4), $\hat{N}$ is no longer preserved (there is, however, a $\mathbb{Z}_2$ parity symmetry). The continuous symmetry resulting from the conserved excitation number is reflected in the sombrero shape of $E_{\text{scJC}}^{(-)}(x, p)$, and thereby in the corresponding CI. We may still achieve a CI even without imposing the RWA if we consider a *bi-modal* quantum Rabi model,

$$\hat{H}_{\text{2mRabi}} = \hbar\omega \left(\frac{\hat{p}_x^2 + \hat{p}_y^2}{2} + \frac{\hat{x}^2 + \hat{y}^2}{2} \right) + \frac{\hbar\Omega}{2}\hat{\sigma}_z + \frac{\hbar g}{\sqrt{2}} \left(\hat{x}\hat{\sigma}_x + \hat{y}\hat{\sigma}_y \right). \tag{6.11}$$

Here, we have assumed that the two photon modes share the same frequency ω and the same atom–light coupling strength g. However, the second mode, denoted by the quadratures $\hat{y}$ and $\hat{p}_y$, couples to the atom with a different phase—it couples to the dipole operator $\hat{\sigma}_y$. This is the familiar $E \times \epsilon$ JT model discussed in Sect. 3.3.2, with a Zeeman field term that breaks time-reversal symmetry.

As we saw in Sect. 3.4, even though the CI is manifested in real space and not in momentum space as for Rashba spin–orbit couplings discussed in Sect. 4.2.2, the presence of it and the accompanying synthetic gauge curvature implies anomalous dynamics in terms of a Hall effect. Contrary to the intrinsic spin Hall effects studied in the aforementioned sections, the breaking of time-reversal symmetry imposed by the 'Zeeman term' $\hbar\Omega\hat{\sigma}_z/2$ leads to an *intrinsic anomalous Hall effect*. The Zeeman splitting causes a population imbalance between the two spin states and in addition to a transverse spin current one also finds a transverse charge current (i.e. a net flow of particles).

A feature of cavity QED settings is that photons are inevitably lost through the cavity mirrors at some rate κ. This leads to unwanted decoherence, but it also provides a handle to gain information about the system without performing destructive measurements—the photons leaking out of the cavity is measured and one obtains, e.g. information about the photon number inside the cavity. We now investigate how the intrinsic anomalous Hall effect manifests in the cavity scenario [4].

With $g = 0$, the Hamiltonian in (6.11) describes two identical uncoupled harmonic oscillators, which implies that any Gaussian state remains Gaussian throughout time-evolution. In particular, a coherent state in either mode would stay coherent. For example, the initial product state $|\alpha_0\rangle|0\rangle|-\rangle$ (with $\alpha_0 \in \mathbb{R}$ and the three kets representing the x-mode, the y-mode, and the atom, respectively) turns into $|\alpha(t)\rangle|0\rangle|-\rangle$ with $\alpha(t) = \alpha_0 e^{-i\omega t}$. The phase space trajectories for the two modes are $(\langle\hat{x}\rangle, \langle\hat{p}_x\rangle) = (\alpha_0\cos(\omega t), \alpha_0\sin(\omega t))$ and $(\langle\hat{y}\rangle, \langle\hat{p}_y\rangle) = (0, 0)$, or $(\langle\hat{x}\rangle, \langle\hat{y}\rangle) = (\alpha_0\cos(\omega t), 0)$. When the light–matter coupling is turned on, i.e. when $g \neq 0$, a transverse synthetic Lorentz force builds up and the trajectory of the Gaussian starts to bend in the harmonic potential $V(x, y)$ (cf. the discussion in Sect. 3.4). This implies that the second mode gets populated. The corresponding trajectories are shown in Fig. 6.2b, while in (a) the photon numbers n_x and n_y of the two modes are presented. An almost perfect swapping of the photons between the two modes is found, which can be ascribed the anomalous Hall effect. The parameters used for the numerical simulations are taken from an actual experiment [5]. In this experiment, photon losses and spontaneous emission of the system (which was not an atom, but a superconducting quantum dot) are inevitably present and to delineate the effect of them, we include a simulation where they have been taken into account, see Fig. 6.2c and grey line in (a). We see the effect of photon dissipation, but the time-scales are such that the anomalous Hall effect still dominates the evolution. The population swapping of photons between the two modes is readily detected experimentally by measuring the photons leaking out from the cavity.

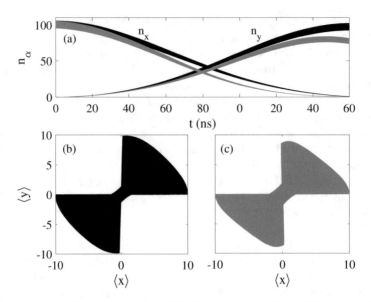

Fig. 6.2 An example of how the anomalous Hall effect reveals in the bi-modal Rabi model (6.11). Initially, while the y-mode is in vacuum, the x-mode is populated by a coherent state with amplitude $\alpha_0 = 10$, i.e. the average number of photons in the two modes are $n_y(0) = 0$ and $n_x(0) = 100$, respectively. The atom starts in the lower state $|-\rangle$. The Hall effect implies a population transfer of photons from the x-mode to the y-mode as demonstrated in (**a**) that shows the populations in the two modes (black lines) vs. time. In (**b**), the trajectory $(\langle x \rangle, \langle y \rangle)$ is shown. For short times, we see a harmonic oscillation given by the time-scale ω^{-1}, but the trajectory slowly builds up a transverse motion meaning that the y-mode gets populated. Finally, after $t \approx 60$ ns all the photons reside in the y-mode. For this plot, we have used experimental parameters [5] $\omega/2\pi = 5.7$ GHz, $\Omega/2\pi = 6.9$ GHz, $g/2\pi = 150$ MHz, spontaneous emission of the atom $\gamma/2\pi = 1.9$ MHz, and photon decay rate $\kappa/2\pi = 250$ kHz. The losses of the atom and the photons are taken into account for the grey curves in (**a**) and (**c**). As is clear, despite the loss rates the anomalous Hall effect survives

6.2 Trapped Ions

Trapped ion systems, like cavity QED, is another setting in which a two-level system is coherently coupled to a harmonic oscillator [6]. If we disregard the internal ionic states, a single trapped ion is a good approximation given by a three-dimensional harmonic oscillator with characteristic trapping frequencies $(\omega_x, \omega_y, \omega_z)$.[1] The idea is to couple internal ionic electronic states to the vibrational motion of the ion by illuminating it with a laser with some frequency ω_L and wave number $\mathbf{k}_L$. For now, let us assume that $\omega_y, \omega_z \gg \omega_x$ and that we impose the two-level approximation on the ion's internal electronic states. The large squeezing of the trap in the y- and

---

[1]For highly excited vibrational states, the traps include anharmonic corrections, but as long as the phonon number is below $\sim$100 this is not a problem.

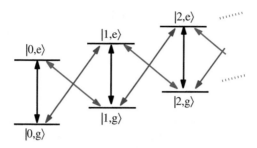

Fig. 6.3 The various transitions in the trapped ion setting: fundamental transitions in black arrows and sideband transitions in red and blue. The coupled states are included in the figure, where $|n, j\rangle$ means a product state of a n-phonon Fock state and the ion in the state $|j\rangle$ with $j = g, e$. The transitions involve different frequencies and are addressed by controlling the frequency of the applied laser light. When driving the fundamental transition the laser photons are (quasi) resonant with the internal ionic transition $|g\rangle \leftrightarrow |e\rangle$, and no change in the ion's motional state occurs. The red and blue sideband transitions, on the other hand, change the phonon number by ∓ 1

z-directions implies that we may assume the ion remains in its vibrational ground state in these directions, i.e. we consider a one-dimensional problem with motion only along the x-direction. Without the laser, the *bare* Hamiltonian of the ion is

$$\hat{H}_{\text{Ion}} = \hbar \omega_x \hat{a}^\dagger \hat{a} + \frac{\hbar \Omega}{2} \hat{\sigma}_z, \qquad (6.12)$$

where we have neglected the constant zero temperature fluctuations $\hbar \omega_x/2$ and Ω is the transition frequency between the ion's internal two electronic ground and excited states $|g\rangle$ and $|e\rangle$, respectively. The Hamiltonian is diagonal in the *bare basis* $\{|n, g\rangle, |n, e\rangle\}$, where $\hat{a}^\dagger \hat{a}|n, j\rangle = n|n, j\rangle$, $j = g, e$, $\hat{\sigma}_z|n, g\rangle = -|n, g\rangle$, and $\hat{\sigma}_z|n, e\rangle = |n, e\rangle$. Thus, the spectrum $E_{n,\pm} = \hbar \omega_x n \pm \frac{\hbar \Omega}{2}$ comprises two ladders separated in energy by $\hbar \Omega$, see Fig. 6.3.

As is seen in Fig. 6.3, transitions between the bare states involve different frequencies. If we restrict the analysis to transitions that shifts the phonon number by at most ± 1, we find three transition frequencies (assuming that $\Omega > \omega_x$); the *fundamental* $\omega_f = \Omega$, the *red sideband* $\omega_{\text{red}} = \Omega - \omega_x$, and the *blue sideband* $\omega_{\text{blue}} = \Omega + \omega_x$. If the frequency ω_L of the laser is chosen to be quasi-resonant with one of these three transitions (and if the laser light wavelength λ_L is long in comparison to the trap size—the *Lamb–Dicke regime*), we may neglect any other transitions apart from those that are quasi-resonant. The red and blue sideband transitions are described by the Jaynes–Cummings Hamiltonian

$$\hat{H}_{\text{red}} = \frac{\hbar \delta_{\text{red}}}{2} \hat{\sigma}_z + \hbar g \left(\hat{\sigma}^+ \hat{a} e^{i\varphi} + \hat{a}^\dagger \hat{\sigma}^- e^{-i\varphi} \right) \qquad (6.13)$$

and the anti-Jaynes–Cummings Hamiltonian

$$\hat{H}_{\text{blue}} = \frac{\hbar \delta_{\text{blue}}}{2} \hat{\sigma}_z + \hbar g \left(\hat{\sigma}^+ \hat{a}^\dagger e^{i\varphi} + \hat{a} \hat{\sigma}^- e^{-i\varphi} \right), \qquad (6.14)$$

respectively. Here, δ_{red} and δ_{blue} are the corresponding detunings between the bare transition frequency and the laser light frequency, and φ is the phase of the laser. By combining these two, we obtain the quantum Rabi mode (cf. (6.4))

$$\hat{H}_{\text{ion}} = \hbar\omega_x \hat{a}^\dagger \hat{a} + \frac{\hbar\Omega}{2}\hat{\sigma}_z + \hbar g \hat{\sigma}_x \left(\hat{a} + \hat{a}^\dagger\right). \tag{6.15}$$

For a full derivation of the Hamiltonian and the expression for the coupling g, we refer to [6]. Equivalent to cavity QED, we see that a transition of the internal ionic state is accompanied by a transition of the oscillator state. The phase φ of the laser can also be adjusted so that the coupling between the ion's internal states and the external vibrational states can be described by a $\hat{\sigma}_y$ term instead of the $\hat{\sigma}_x$ term in the above Hamiltonian [7].

The derivation above may be straightforwardly generalised to the two- and three-dimensional cases by relaxing the squeezing of the trap in the y- and z-direction and by including one or two lasers that 'dress' the ion also in these directions. By controlling the corresponding phases we find the $E \times \epsilon$ and the $T \times \epsilon$ JT Hamiltonians

$$\hat{H}_{E\times\epsilon} = \hbar\omega\left(\frac{\hat{p}_x^2}{2} + \frac{\hat{p}_y^2}{2} + \frac{\hat{x}^2}{2} + \frac{\hat{y}^2}{2}\right) + \hbar g\sqrt{2}\left(\hat{x}\hat{\sigma}_x + \hat{y}\hat{\sigma}_y\right),$$

$$\hat{H}_{T\times\epsilon} = \hbar\omega\left(\frac{\hat{p}_x^2}{2} + \frac{\hat{p}_y^2}{2} + \frac{\hat{p}_z^2}{2} + \frac{\hat{x}^2}{2} + \frac{\hat{y}^2}{2} + \frac{\hat{z}^2}{2}\right)$$

$$+ \hbar g\sqrt{2}\left(\hat{x}\hat{\sigma}_x + \hat{y}\hat{\sigma}_y + \hat{z}\hat{\sigma}_z\right), \tag{6.16}$$

where we let all the trapping frequencies to be the same, e.g. $\omega_x = \omega_y = \omega_z = \omega$, and the coupling g the same in all directions.

If many ions are held in the trap, they tend to repel each other due to their electric charge. If the trap is squeezed in two directions the ions tend to organise along a chain in the third direction. Such an ion chain was considered in [7], and furthermore the ions were also coupled according to the $E \times \epsilon$ JT model in the transverse directions. Since the ions interact via Coulomb interaction this setup realises a 'many-body' JT model. The corresponding JT effect, i.e. the formation of a sombrero shaped lower APS, was identified as a continuous phase transition into a phonon condensation.

6.3 Classical Optics

When light is propagating through a medium it experiences a refracting index n, which affects the properties of the light. From Maxwell's equations we can derive a wave equation for the electric field,

$$\nabla^2 E - \frac{n^2}{c^2}\frac{\partial^2 E}{\partial t^2} = 0, \tag{6.17}$$

where $E = E(x, y, z, t)$ is the field amplitude and c is the speed of light. In (6.17), we have assumed that there is no coupling between the different polarisation directions of the electric field. This consequently means that we assume the change in refractive index is small over a length scale given by the wavelength of the light. If we consider a monochromatic light field with

$$E(x, y, z, t) = \varepsilon(x, y, z)e^{-i\omega t}, \tag{6.18}$$

where ω is the frequency of the light, we obtain the Helmholtz equation

$$\nabla^2 \varepsilon - (n_0 + n_1)^2 k^2 \varepsilon = 0 \tag{6.19}$$

with $k = \omega/c$ and n expressed in terms of a background refractive index n_0 plus a small, possibly space-dependent, refractive index n_1. In the following, we consider a light beam propagating in the z-direction. We therefore seek solutions of the form

$$\varepsilon(x, y, z) = \xi(x, y, z)e^{-in_0 kz}. \tag{6.20}$$

For a light beam propagating in the z-direction it is natural to assume that changes in ξ and $\partial \xi/\partial z$ within a distance of the wavelength λ are negligible. In other words, we require that

$$|\frac{\partial \xi}{\partial z}| \ll k|\xi|,$$

$$|\frac{\partial^2 \xi}{\partial^2 z}| \ll k|\frac{\partial \xi}{\partial z}|, \tag{6.21}$$

where $k = 2\pi/\lambda$. This also means that we assume that $\xi(x, y, z)$ varies approximately as e^{ikz} over a distance of a few wavelengths λ. The Helmholtz equation still needs to be fulfilled with

$$\left(\frac{\partial^2}{\partial x^2} + \frac{\partial^2}{\partial y^2} + \frac{\partial^2}{\partial z^2}\right)\xi e^{in_0 kz} + (n_0 + n_1)^2 k^2 \xi e^{in_0 kz} = 0. \tag{6.22}$$

Using the assumptions in (6.21), we find

$$\frac{\partial^2}{\partial z^2}\xi e^{in_0 kz} = \left(\frac{\partial^2 \xi}{\partial z^2} + 2in_0 k\frac{\partial \xi}{\partial z} - n_0^2 k^2 \xi\right)e^{in_0 kz}$$

$$\approx (2in_0 k\frac{\partial \xi}{\partial z} - n_0 k^2 \xi)e^{in_0 kz}. \tag{6.23}$$

The Helmholtz equation (6.22) then becomes

$$\left(\frac{\partial^2}{\partial x^2} + \frac{\partial^2}{\partial y^2}\right)\xi + 2in_0k\frac{\partial\xi}{\partial z} + (n^2 - n_0^2)k^2\xi = 0. \tag{6.24}$$

This is called the paraxial wave equation and is in the form of a Schrödinger equation for a particle in the two-dimensional (x, y)-plane and where z is interpreted as time t. Explicitly, this translates into the equation

$$i\frac{\partial}{\partial z}\xi = \left(-\frac{1}{2n_0k}\nabla_{xy}^2 - kn_1(x, y, z)\right)\xi. \tag{6.25}$$

The spatially dependent refractive index n with its small deviation n_1 from the background n_0, can consequently act as an effective potential for the light. This potential can guide and trap the light if we can locally change the refractive index. We can therefore envisage an array of wave guides in a piece of glass where the refractive index has been changed locally forming the wave guides. This opens up a powerful tool to study two-dimensional dynamics of discrete coherent light in the transversal plane of the light beam.

The required change in refractive index can be achieved by laser writing [8], where a focal point of a femtosecond laser beam is traversing through a piece of glass, and thereby locally changing the refractive index. With this technique, it is in principle possible to create any two-dimensional lattice structure in the transversal plane. For a lattice structure with weak evanescent coupling between the lattice sites, a tight-binding model can be used, with which we can calculate the band structure of the system. We can therefore emulate lattice models known from solid state physics, but now using light which can tunnel from lattice site to lattice site. Dirac cones can readily be obtained in the dispersion relation provided that we choose the appropriate lattice structure. As an example, we consider an edge-centred square lattice, also called a Lieb lattice [9], as shown in Fig. 6.4. We obtain the energy spectrum by Fourier transforming the corresponding Hamiltonian, which results in three bands,

$$\Omega_\pm(\mathbf{q}) = \pm 2\sqrt{\kappa_x^2\cos^2(q_xa) + \kappa_y^2\sin^2(q_ya)},$$
$$\Omega_0(\mathbf{q}) = 0, \tag{6.26}$$

where κ_x and κ_y are the coupling constants for nearest neighbour sites along the x and y axes, and a is the lattice constant.

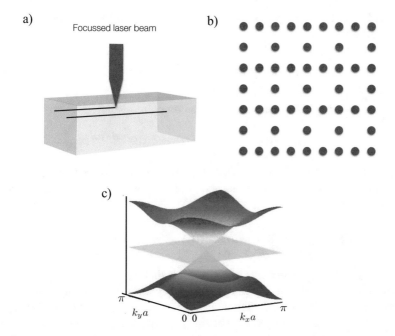

Fig. 6.4 (**a**) The focussed femtosecond laser beam creates a local change in the refractive index which can provide a wave guide for the light. (**b**) The edge-centred square lattice, also called the Lieb lattice, is one example of a non-trivial two-dimensional lattice in the transversal plane of the laser beam. (**c**) The Lieb lattice has three bands, including one flat band. Here, the coupling strengths between the lattice sites are chosen to be the same in x- and y-direction

6.4 Open Quantum Systems

In this section, we describe how a new type of CIs, called *exceptional points* (EPs), may appear in open quantum systems. Such systems are, in a strict sense, any system that is in contact with its surrounding. However, we do not think of a driven system, e.g. a particle whose evolution is governed by an explicitly time-dependent Hamiltonian, as open. In principle, the surrounding need not be large in terms of degrees of freedoms; it can be any other quantum system. However, one should keep in mind that the surrounding is some sort of environment that the system only couples weakly to. Regardless of the situation, any open quantum system should be described by time-evolution that is non-unitary.

Let us start by an example that brings us to the *Lindblad master equation* in the next subsection. Consider a harmonic oscillator, e.g. a single mode of the radiation field, that is in contact with a zero temperature environment. Photons may decay with some rate κ into the environment, and in the long run the oscillator should be in a thermal state, which for the zero temperature environment implies the vacuum state. A first attempt to describe such a situation would be to add a complex term to the Hamiltonian

$$\hat{H}_{\text{eff}} = \hbar\omega \left(\hat{a}^{\dagger}\hat{a} + \frac{1}{2} \right) - i\kappa \hat{a}^{\dagger}\hat{a}. \tag{6.27}$$

The corresponding Heisenberg equations become

$$\partial_t \hat{a} = -i(\omega - i\kappa)\hat{a} \quad \Rightarrow \quad \hat{a}(t) = \hat{a}(0)e^{-\kappa t - i\omega t}, \tag{6.28}$$

which further implies $\hat{a}^{\dagger}(t) = \hat{a}^{\dagger}(0)e^{-\kappa t - i\omega t}$. The time evolved operators should obey bosonic commutation relations at any instant of time, at $t = 0$ as well as for $t \rightarrow \infty$, but we find

$$\lim_{t \rightarrow \infty} \left[\hat{a}(t), \hat{a}^{\dagger}(t) \right] = \lim_{t \rightarrow \infty} \left[\hat{a}(0), \hat{a}^{\dagger}(0) \right] e^{-2\kappa t} = \lim_{t \rightarrow \infty} e^{-2\kappa t} = 0, \tag{6.29}$$

which clearly cannot be correct! The mistake is that we have not taken fluctuations into account; according to the *fluctuation–dissipation theorem* any dissipation is accompanied by fluctuations. The more correct description would be to replace the Heisenberg equations by the *Heisenberg–Langevin* equations, which include a stochastic term that accounts for the fluctuations arising due to the coupling to the environment [10].

6.4.1 The Lindblad Master Equation

In the example above, we showed that we end up with unphysical results if dissipation is taken into account by simply adding a complex term to the Hamiltonian. This was done in the Heisenberg picture, where the operators evolve in time. Of course, we come to a contradiction also if we work in the Schrödinger picture. A state would then evolve as $\Psi(t) = \exp\left(-i\hat{H}_{\text{eff}}t\right)\Psi(0)$, and clearly the norm is not preserved since $\lim_{t \rightarrow \infty} \Psi(t) = 0$. We already said that the resolution to this problem is the environment-induced fluctuations. To consider fluctuations it is necessary to introduce density matrices—fluctuations demolish information in the phases of the quantum states and the states do not remain pure.

The idea of the approach is outlined in Fig. 6.5. Some initial state $\hat{\rho}(0)$ evolves under some Hamiltonian

$$\hat{H}_{\text{tot}} = \hat{H}_{\text{S}} + \hat{H}_{\text{E}} + \hat{H}_{\text{SE}}, \tag{6.30}$$

where $\hat{H}_{\text{S}}$ is the system Hamiltonian, $\hat{H}_{\text{E}}$ the environment Hamiltonian, and $\hat{H}_{\text{SE}}$ is describing the interaction between the two. The system state at the initial and final times is $\hat{\rho}_{\text{S}}(0) = \text{Tr}_E\left[\hat{\rho}(0)\right]$ and $\hat{\rho}_{\text{S}}(t) = \text{Tr}_E\left[\hat{\rho}(t)\right]$, respectively, where

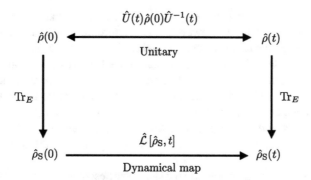

Fig. 6.5 Schematic of the time-evolution of the system + environment. At time $t = 0$ the system plus environment is in some general state $\hat{\rho}(0)$, and by tracing over the environment (bath) degrees-of-freedom, the reduced state $\hat{\rho}_S(0)$ for the system alone is obtained. The state of the full system evolves unitarily into the state $\hat{\rho}(t)$ for any later time t, while the system state $\hat{\rho}_S(t)$ at that time is again given by the partial trace over the environment. The goal is to find an equation (dynamical map) rendering the system's time evolution; $\hat{\rho}_S(0) \rightarrow \hat{\rho}_S(t)$. Such a map is not reversible, just like the partial traces. This irreversibility is marked by the directions of the arrows

$\hat{\rho}(t) = \hat{U}(t)\hat{\rho}(0)\hat{U}^\dagger(t)$ is the time evolved full state (i.e. $\hat{U}(t) = \exp\left(-i\hat{H}_{\text{tot}}t\right)$, $\hbar = 1$), and the partial trace is over the environment degrees of freedom. A *dynamical map*, or *Liouvillian* $\hat{\mathscr{L}}$, takes the initial system state $\hat{\rho}_S(0)$ to the final system state $\hat{\rho}_S(t)$. Note that $\hat{\mathscr{L}}$ is a *super-operator*, meaning that it acts on an operator $\hat{\rho}(0)$ and returns another operator $\hat{\rho}(t)$. The problem is to find the equation that results in the corresponding dynamical map. If we aim at finding an equation that involves only system degrees of freedom, this is in general not possible unless we impose some approximations.

Lindblad proved [11] that the most general Markovian quantum master equation is on the form

$$\partial_t \hat{\rho}(t) = \hat{\mathscr{L}}\left[\hat{\rho}(t)\right] \equiv -\frac{i}{\hbar}\left[\hat{\rho}(t), \hat{H}\right] + \sum_i \kappa_i \left(2\hat{L}_i \hat{\rho}(t)\hat{L}_i^\dagger - \hat{L}_i^\dagger \hat{\rho}(t) - \hat{\rho}(t)\hat{L}_i^\dagger \hat{L}_i\right),$$

(6.31)

where we have defined the Liouvillian $\hat{\mathscr{L}}$ and we have omitted the system subscript S on the density operator. What we mean by most general is that it is the most general form of such an equation that preserves the physical properties of the density operator, i.e. the trace (norm) of $\hat{\rho}(t)$ is intact, and the density operator remains positive semi-definite and Hermitian. Usually such a dynamical map is referred to as completely positive trace preserving (CPTP). The first part of the equation accounts for the unitary evolution generated by some Hamiltonian $\hat{H}$. The second part arises from the interaction with the environment. $\hat{L}_i$ are called *Lindblad jump operators*, and if the dimension of the Hilbert space is d there are $d^2 - 1$ linearly independent operators $\hat{L}_i$ such that the sum may be restricted to only those. For a two-level system, e.g. we can take the Pauli matrices as a set of Lindblad jump operators. However, it is often more practical to consider non-Hermitian $\hat{L}_i$'s, which describe creation and annihilation of particles or excitations. For example, to describe spontaneous emission of a two-level atom into a zero temperature environment one would restrict the Lindblad operators to only one, namely, $\hat{L}_1 = \hat{\sigma}^-$. While for photon decay out of a cavity into a zero temperature environment one typically has $\hat{L}_1 = \hat{a}$. This explains why the $\hat{L}_i$'s are called 'jump operators'; application $\hat{L}_i \hat{\rho}(t) \hat{L}_i^\dagger$ implies that the excitation or particle number is decreased by one unit—the state makes a jump into a new state with lower excitations.

The *Lindblad master equation* (6.31) is manifestly Markovian, meaning that any information leaking from the system to the environment is forever lost. Thus, we can view the environment as continuously measuring the system, and as it gains information the state of the system cannot in general remain pure. Without a Markovian approximation, it would mean that the system would become entangled with the environment. In the derivation of the Lindblad master equation starting from some specific system, two more approximations are imposed. The *Born approximation* assumes that the environment is large on the system scale, such that the environment is negligibly influenced by the coupling to the system, i.e. the environment state (typically a thermal state) stays unaffected by the system. The other is the *secular approximation*, which is the open system analogue to the RWA introduced in Sect. 6.1.1. Here, the secular approximation accounts for omitting interaction terms where a particle can be created or destroyed simultaneously in the system and the environment. We do not present any derivation of the equation here, but refer to any book on quantum optics, see, for example, [10].

Let us return to the problem we started with—the dissipative harmonic oscillator. For a zero temperature thermal environment of harmonic oscillators, one derives the Lindblad master equation [10]

$$\partial_t \hat{\rho}(t) = -i \left[\hat{\rho}(t), \omega \hat{a}^\dagger \hat{a} \right] + \kappa \left(2\hat{a}\hat{\rho}(t)\hat{a}^\dagger - \hat{a}^\dagger \hat{a} \hat{\rho}(t) - \hat{\rho}\hat{a}^\dagger \hat{a} \right). \tag{6.32}$$

The equation has a unique steady state that is the vacuum, i.e. $\lim_{t\to\infty} \hat{\rho}(t) = |0\rangle\langle 0|$. Naturally, this steady state is reached regardless of the initial state, which means that the evolution must be irreversible. For an initial coherent state $\hat{\rho}(0) = |\alpha\rangle\langle\alpha|$ the time evolved state remains coherent (i.e. it is a pure state throughout the evolution),

$$\hat{\rho}(t) = |\alpha(t)\rangle\langle\alpha(t)|, \qquad \alpha(t) = \alpha e^{-i\omega t - \kappa t}. \tag{6.33}$$

To see how decoherence sets in we start from a *cat state*

$$\hat{\rho}(0) = \frac{1}{N} \left(|\alpha\rangle\langle\alpha| + |-\alpha\rangle\langle-\alpha| + |\alpha\rangle\langle-\alpha| + |-\alpha\rangle\langle\alpha| \right), \tag{6.34}$$

where $N = 2(1 + \mathrm{Re}(\langle \alpha | - \alpha \rangle))$. The time evolved state is given by [12]

$$\hat{\rho}(t) = \frac{1}{N(t)} \big(|\alpha(t)\rangle \langle \alpha(t)| + |-\alpha(t)\rangle \langle -\alpha(t)|$$

$$+ C(t)|\alpha(t)\rangle \langle -\alpha(t)| + C(t)|-\alpha(t)\rangle \langle \alpha(t)| \big), \qquad (6.35)$$

where the factor multiplying the coherence terms reads

$$C(t) = \exp\left(-2|\alpha|^2(1 - e^{-2\kappa t})\right) \qquad (6.36)$$

if we assume α to be real. For $\kappa t \ll 1$, we have that the coherence terms vanish as $C(t) \approx e^{-2\kappa |\alpha|^2 t}$. By noting that $2|\alpha|^2$ is the 'distance' between the two coherent states that form the cat state we see how the cat becomes extremely fragile to decoherence the larger the cat is (the cat state size is the number of bosons $n = |\alpha|^2$). Thus, after an exponentially short time the cat has turned into a statistical mixture between the two coherent states.

6.4.2 Exceptional Points

A CI is characterised by crossing at a single point of eigenvalues of a parameter-dependent Hamiltonian. The Hermiticity of the Hamiltonian implies that the energies are real-valued; a natural condition since energy is supposedly a physical quantity. However, it turns out that in effective theories, the matrices generating time-evolution need not be Hermitian. As we will see, the Liouvillian of a Lindblad master equation can typically not be represented by a Hermitian matrix. The eigenvalues of a non-Hermitian matrix are normally complex-valued—the imaginary part represents some sort of relaxation. A CI in this setting is taken if both the real and imaginary parts of the eigenvalues coincide at one and the same point in the parameter space [13]. Such EPs are similar to degeneracies for Hermitian matrices, but still not exactly the same.

Let us first consider an example. The Landau–Zener model is one of few time-dependent models that enables an analytical solution [14, 15]. The Schrödinger equation takes the form (we put $\hbar = 1$ from now on)

$$i \frac{\partial}{\partial t} \begin{bmatrix} c_1(t) \\ c_2(t) \end{bmatrix} = \hat{H}_{\mathrm{LZ}}(t) \begin{bmatrix} c_1(t) \\ c_2(t) \end{bmatrix} = \begin{bmatrix} \lambda t & g \\ g & -\lambda t \end{bmatrix} \begin{bmatrix} c_1(t) \\ c_2(t) \end{bmatrix}, \qquad (6.37)$$

such that a solution is given by $|\Psi(t)\rangle = c_1(t)|1\rangle + c_2(t)|2\rangle$. Here, the real parameter g is the coupling between the two states and assumed constant in time, while the diagonal terms are the energies of the diabatic states $|1\rangle$ and $|2\rangle$ in the absence of any coupling, and they are linearly quenched in time with a rate $\pm\lambda$. At $t = 0$ the two diabatic energies $\pm\lambda t$ cross. However, due to the coupling between the two states the crossing becomes avoided. Provided that we initialise the system in one of the diabatic states $|1\rangle$ or $|2\rangle$ at $t = -\infty$, we seek the probability to remain in that state at $t = +\infty$. The Landau–Zener formula gives this probability as

$$P_{LZ} = \exp\left(-\pi\frac{g^2}{\lambda}\right). \tag{6.38}$$

The argument $\Gamma = \pi g^2/\lambda$ of the exponent can be seen as an 'adiabaticity parameter'; the larger it is, the more adiabatic is the process. Thus, a larger coupling g or a smaller rate λ provide an adiabatic transition. In the adiabatic limit the transition probability is unity corresponding to a perfect swap of the diabatic states, i.e. $|1\rangle \to |2\rangle$. In principle the amplitudes $c_1(t)$ and $c_2(t)$ can be found analytically for any times, but the expressions are rather complicated. We may note, though, that in the fully adiabatic sweep from $t = -\infty$ to $t = +\infty$, we swap the diabatic states $|1\rangle \leftrightarrow |2\rangle$. However, a closer look shows that we also acquire an additional sign change of the state, so to be precise we have $|1\rangle \leftrightarrow -|2\rangle$.

Time t is not a proper parameter, like g or λ, but nevertheless let us think of it as an external parameter in the adiabatic limit. What happens when we analytically continue the time out in the complex plane, $t \to t_r + it_i$? Thus, we extend the Landau–Zener Hamiltonian into the complex plane $\hat{H}_{LZ}(t_R + it_i)$. The corresponding eigenvalues are

$$\epsilon_\pm = \pm\sqrt{g^2 + \lambda^2(t_r + it_i)^2}. \tag{6.39}$$

By letting $t_i = 0$, we regain the adiabatic energies of the Landau–Zener model with the characteristic avoided crossing at $t = 0$ and with an energy gap $2g$ at that instant. For $(t_r, t_i) = (0, \pm g/\lambda)$ we have that $\epsilon_- = \epsilon_+ = 0$—the two eigenvalues become identical. This marks an EP. In Fig. 6.6, we show the real and imaginary parts of the eigenvalues and it is clear how the EPs manifest. If we follow the imaginary axis, $t_r = 0$, say in the negative direction we have that at the origin the imaginary part of the eigenvalue is zero while there is a $2g$ gap between the real parts. As we move along the imaginary axis, the imaginary parts stay zero, while the gap between the real parts close like a square root singularity. At the EP the gap is fully closed and instead the imaginary parts opens up a gap.

Returning to the actual Landau–Zener model, we saw that if we drive the system adiabatically the diabatic states will be swapped by the quench. That is, the system follows adiabatically the instantaneous eigenstates, i.e. the adiabatic states will not be swapped by the Landau–Zener sweep. We assume that the parameters are such that we are in the adiabatic regime. The system is integrated from $t = -\infty$ to $t = +\infty$, which in the complex time plane means that the integration is along the real time axis. When we close the integration curve at infinity, we see that we must encircle one of the EPs. We know that when encircling a CI adiabatically we pick up a π overall phase resulting in a sign change of the wave function. Here is a crucial

(a) $\mathrm{Re}(\epsilon_{\pm})$ (b) $\mathrm{Im}(\epsilon_{\pm})$

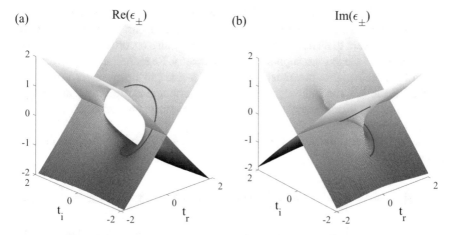

Fig. 6.6 The adiabatic Landau–Zener energies (6.39) in the complex time plane. (**a**) Along the real time-axis ($t_i = 0$), the imaginary part of the energy vanishes and the real part displays the Landau–Zener avoided crossing. (**b**) On the other hand, along the imaginary time-axis ($t_r = 0$), the energies possess EPs at $t_i = \pm 1$. At the EP both the real and imaginary parts of the two energies become degenerate in a square root singularity. Upon encircling one such EP, indicated by the thick red curves, the state of the system is swapped. In this plot, we use dimensionless parameters with $\hbar = 1$, and $g = \lambda = 1$

difference between an EP and a CI; when encircling adiabatically an exception point, we swap the states. This is visualised in Fig. 6.6 by the red thick lines—the surfaces form two Riemann sheets and you have to encircle the EP twice to come back to the same sheet. But this is what we saw in the Landau–Zener adiabatic sweep; we went from one diabatic state to the other.

Put in general terms, at the EP we have $|\epsilon_+ - \epsilon_-| = 0$, equivalently to a degeneracy for Hermitian Hamiltonians. However, when we adiabatically encircle such a point repeatedly, the state obeys, for example,

$$|1\rangle \to -|2\rangle \to -|1\rangle \to |2\rangle \to |1\rangle. \tag{6.40}$$

Thus, the EP-encircling can be represented by the unitary operator $\hat{U} = \exp\left(-\pi \hat{\sigma}_y / 2\right)$. Note that after two revolutions we are back with an overall minus sign, and we need four revolutions to return back to the initial state. Another property of the EPs is that at these points the corresponding eigenvectors coalesce such that there exist only a single eigenvector. Thus, at the EPs the matrix is not diagonalisable. Instead there exists a similarity matrix $\hat{S}$ that casts the matrix into *Jordan block form*, i.e. it has the structure

$$\hat{H} = \begin{bmatrix} \epsilon & 1 \\ 0 & \epsilon \end{bmatrix}. \tag{6.41}$$

Higher order EPs are also possible, where more than two eigenvalues coincide and the Jordan blocks will consequently be of higher dimensions. However, in physical systems they seem to be not very common. EPs may also become lines in parameter space provided the system supports some symmetry. Regarding the eigenvectors of a non-Hermitian matrix one must separate left from right eigenvectors. Apart from at the EP, the left and right eigenvectors are bi-orthogonal meaning that any right eigenvector $|\phi_i\rangle$ is orthogonal to the complex conjugate of any left eigenvector $\langle\phi_j|$; $\langle\phi_j^*|\phi_i\rangle = \delta_{ij}$.

Let us consider one further example; the $E \times \epsilon$ JT model Hamiltonian in the presence of losses,

$$\hat{H}_{E\times\epsilon} = \frac{\hat{p}_x^2 + \hat{p}_y^2}{2m} + \frac{m\omega^2}{2}\left(\hat{x}^2 + \hat{y}^2\right) + g\left(\hat{x}\hat{\sigma}_z + \hat{y}\hat{\sigma}_y\right) - i\gamma\hat{\sigma}_z. \tag{6.42}$$

The last term, describing the loss, makes the Hamiltonian non-Hermitian. Note that the $\hat{x}$ variable couples to $\hat{\sigma}_z$ and not to $\hat{\sigma}_x$ (cf. (6.16)). The APSs are

$$\epsilon_\pm(x, y) = \frac{m\omega^2}{2}\left(\hat{x}^2 + \hat{y}^2\right) \pm \sqrt{(gx - i\gamma)^2 + (gy)^2}. \tag{6.43}$$

We find two EPs for $(x, y) = (0, \pm\gamma/g)$, see Fig. 6.7a, b. Thus, the distance between the two EPs in the xy-plane is $2\gamma/g$, and in the limit of $\gamma \to 0$, the two EPs merge into the CI of the $E \times \epsilon$ JT model. If instead the $\hat{x}$ variable would couple to $\hat{\sigma}_x$ the square root of the adiabatic surfaces (6.43) would read $\sqrt{(gx)^2 + (gy)^2 - \gamma^2}$. We note that the non-Hermitian term does not break the polar symmetry of the model and the EPs form a continuous circle around the origin $(x, y) = (0, 0)$ in this scenario. The formation of such a ring of EPs, emerging from a Dirac CI, has been observed in optical experiments [16].

As we have already argued, when describing the physics by an effective non-Hermitian Hamiltonian, such as (6.42), we neglect fluctuations, and this inevitably leads to non-physical effects. In the Lindblad master equations, fluctuations are taken into account, and in particular any initial physical state will remain a proper physical state for all times. If we were to treat the open JT model in the Lindblad formalism, we should consider the master equation

$$\partial_t\hat{\rho}(t) = i\left[\hat{\rho}(t), \hat{H}_{E\times\epsilon}\right] + \gamma\left(2\hat{\sigma}^-\hat{\rho}(t)\hat{\sigma}^+ - \hat{\sigma}^+\hat{\sigma}^-\hat{\rho}(t) - \hat{\rho}(t)\hat{\sigma}^+\hat{\sigma}^-\right). \tag{6.44}$$

The correspondence of the BOA would be to exclude the kinetic energy from $\hat{H}_{E\times\epsilon}$ in the above master equation. We can rewrite the above equation in terms of the Bloch vector $\mathbf{R} = (R_x, R_y, R_z)$, which is defined as

$$\hat{\rho} = \frac{1}{2}\left(\hat{1} + \mathbf{R}\cdot\hat{\sigma}\right), \tag{6.45}$$

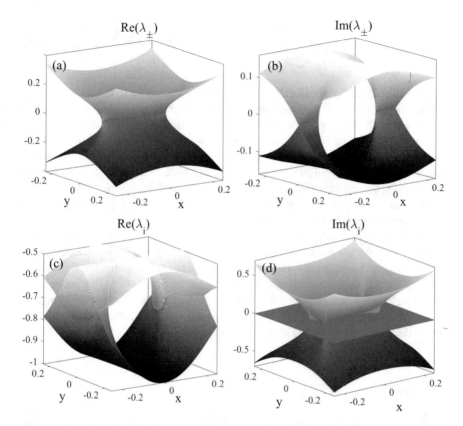

Fig. 6.7 The real and imaginary parts of the APSs for the open $E \times \epsilon$ JT model. The losses appear either as adding an imaginary term $-\gamma\hat{\sigma}_z$ to the Hamiltonian (6.42), (**a**) and (**b**), or adiabatically diagonalising the Liouvillian of the Lindblad master equation (6.46), (**c**) and (**d**). In the upper two plots, the potentials (6.43) are shown and we see a similar structure as in the Landau–Zener problem shown in Fig. 6.6. For the Liouvillian, (**c**) and (**d**), there are three surfaces. We see an evident resemblance between the two models (due to the construction of the master equation (6.46), (**a**) and (**d**), and (**b**) and (**c**) should be compared). However, one difference between the two models is the number of EPs; there are twice as many for the Lindblad master equation. For this plot, we use dimensionless parameters, and to better visualise the structure of the EPs we take $\omega = 0$

with $\hat{\sigma} = (\hat{\sigma}_x, \hat{\sigma}_y, \hat{\sigma}_z)$. By neglecting the kinetic energy term, (6.44) can be expressed in terms of the Bloch vector as

$$\partial_t \begin{bmatrix} R_x(t) \\ R_y(t) \\ R_z(t) \end{bmatrix} = \mathbf{M}_{\mathscr{L}} \begin{bmatrix} R_x(t) \\ R_y(t) \\ R_z(t) \end{bmatrix} + \mathbf{b}_{\mathscr{L}}$$

$$= \begin{bmatrix} -\gamma/2 & -2x & 0 \\ 2x & -\gamma/2 & -2y \\ 0 & 2y & -\gamma \end{bmatrix} \begin{bmatrix} R_x(t) \\ R_y(t) \\ R_z(t) \end{bmatrix} + \begin{bmatrix} 0 \\ 0 \\ 4\gamma \end{bmatrix}. \qquad (6.46)$$

The matrix $\mathbf{M}_{\mathscr{L}}$ is the Liouvillian expressed in the Bloch representation. The term $\mathbf{b}_{\mathscr{L}}$ is sort of a 'pump' that prohibits the null vector $\mathbf{R} = (0, 0, 0)$ to be a trivial steady state. The corresponding adiabatic surfaces for the Lindblad master equation are the eigenvalues of $\mathbf{M}_{\mathscr{L}}$, which we show in Fig. 6.7c, d. Instead of two surfaces, we now have three, and we see similarities between those of the Lindblad equation and those from the non-Hermitian Hamiltonian. Note that the imaginary and real parts of the two sort of surfaces have been interchanged; the Bloch equations (6.46) are defined without the 'i' on the left- hand side. Two additional EPs appear between the surfaces of the Lindblad equation.

Returning to the matrix $\mathbf{M}_{\mathscr{L}}$ of (6.46), we note that for the closed Hamiltonian system ($\gamma = 0$), the matrix $i\mathbf{M}_{\mathscr{L}}$ is Hermitian (i.e. $\mathbf{M}_{\mathscr{L}}$ is anti-symmetric). This is true for any dimension d—the Liouvillian matrix is anti-symmetric for a closed system. The Bloch parametrisation used above can be used for any dimensions d by replacing the Pauli matrices with the *generalised Gell-Mann matrices*. The length of the generalised Bloch vector will be $d^2 - 1$, so that the number of adiabatic surfaces grows quadratically with the Hilbert space dimension.

References

1. Scully, M.O., Zubairy, M.S.: Quantum Optics. Cambridge University Press, Cambridge (1997)
2. Larson, J.: Absence of vacuum induced Berry phases without the rotating wave approximation in cavity QED. Phys. Rev. Lett. **108**, 033601 (2012)
3. Niemczyk, T., Deppe, F., Huebl, H., Menzel, E.P., Hocke, F., Schwarz, M.J., Garcia-Ripoll, J.J., Zueco, D., Hümmer, T., Solano, E., Marx, A., Gross, R.: Circuit quantum electrodynamics in the ultrastrong-coupling regime. Nature Phys. **6**, 772 (2010)
4. Larson, J.: Analog of the spin-orbit-induced anomalous Hall effect with quantized radiation. Phys. Rev. A **81**, 051803(R) (2010)
5. Schuster, D.I., Houck, A.A., Schreier, J.A., Wallraff, A., Gambetta, J.M., Blais, A., Frunzio, L., Majer, J., Johnson, B., Devoret, M.H., Girvin, S.M., Schoelkopf, R.J.: Resolving photon number states in a superconducting circuit. Nature **445**, 515 (2007)
6. Leibfried, D., Blatt, R., Monroe, C., Wineland, D.: Quantum dynamics of single trapped ions. Rev. Mod. Phys. **75**, 281 (2003)
7. Porras, D., Ivanov, P.A., Schmidt-Kaler, F.: Quantum simulation of the cooperative Jahn-Teller transition in 1D ion crystals. Phys. Rev. Lett. **108**, 235701 (2012)
8. Davis, K.M., Miura, K., Sugimoto, N., Hirao, K.: Writing waveguides in glass with a femtosecond laser. Opt. Lett. **21**, 1729 (1996)
9. Mukherjee, S., Spracklen, A., Choudhury, D., Goldman, N., Öhberg, P., Andersson, E., Thomson, R.R.: Observation of a localized flat-band state in a photonic lieb lattice. Phys. Rev. Lett. **114**, 245504 (2015)
10. Gardiner, C., Zoller, P.: The Quantum World of Ultra-Cold Atoms and Light – Book I, Foundations and Quantum Optics. Imperial College Press, London (2014)

11. Lindblad, G.. On the generators of quantum dynamical semigroups. Commun. Math. Phys. **48**, 119 (1976)
12. Walls, D.F., Milburn, G.J.: Effect of dissipation on quantum coherence. Phys. Rev. A **31**, 2403 (1985)
13. Heiss, W.D.: The physics of exceptional points. J. Phys. A: Math. Theor. **45**, 444016 (2012)
14. Landau, L.D.: Zur Theorie der Energieubertragung II. Z. Sowjetunion **2**, 46 (1932)
15. Zener, C.: Non-adiabatic crossing of energy levels. Proc. R. Soc. Lond. A **137**, 696 (1932)
16. Zhen, B., Hsu, C.W., Igarashi, Y., Lu, L., Kaminer, I., Pick, A., Chua, S.-L., Joannopoulos, J.D., M. Soljacic, M.: Spawning rings of exceptional points out of Dirac cones. Nature **525**, 354 (2015)

Appendix A
Identical Particles

A.1 Second Quantisation

The methods and ideas behind second quantisation can be found in numerous textbooks, see, e.g., [1]. In this appendix, we follow a less standard, and also less rigorous, approach which still gives the general picture [2]. When dealing with N indistinguishable particles, we have that the many-body wave function

$$\Psi(\mathbf{r}_1, \ldots, \mathbf{r}_N, t) \tag{A.1}$$

switches sign or remain intact upon swapping any two coordinates $\mathbf{r}_i$ and $\mathbf{r}_j$. This is determined by whether the particles are fermions or bosons, respectively. Given the wave function, the swapping of coordinates leaves the particle density $P(\mathbf{r}_1, \ldots, \mathbf{r}_N, t) = |\Psi(\mathbf{r}_1, \ldots, \mathbf{r}_N, t)|^2$ unchanged. The wave function is a complex function living in a $3N$-dimensional configuration space. We are, however, more interested in knowing the particle density in the real three-dimensional space.

If we start by considering a single particle, we can expand the wave function in some time-independent orthonormal basis,

$$\Psi(\mathbf{r}, t) = \sum_n a_n(t)\phi_n(\mathbf{r}) \quad \Leftrightarrow \quad |\Psi(t)\rangle = \sum_n a_n(t)|\phi_n\rangle. \tag{A.2}$$

The Schrödinger equation in the given basis becomes

$$i\hbar \frac{da_n(t)}{dt} = \sum_m \langle\phi_n|\hat{H}|\phi_m\rangle a_m(t), \tag{A.3}$$

© Springer Nature Switzerland AG 2020
J. Larson et al., *Conical Intersections in Physics*, Lecture Notes in Physics 965,
https://doi.org/10.1007/978-3-030-34882-3

which alternatively can be written as

$$\frac{da_n(t)}{dt} = \frac{\partial\langle\hat{H}\rangle}{\partial(i\hbar a_n^*)},$$

$$\frac{d(i\hbar a_n^*(t))}{dt} = -\frac{\partial\langle\hat{H}\rangle}{\partial a_n} \tag{A.4}$$

with the energy expectation value $\langle\hat{H}\rangle = \langle\psi|\hat{H}|\psi\rangle$. Equation (A.4) has the same form as Hamilton's equations for the canonical variables x and p. The physical interpretation of the amplitudes a_n is that $|a_n|^2$ is the probability to find the particle in state $|\phi_n\rangle$ in a measurement. If we would repeat the experiment N times, we have that the number of times we would find the system in the particular state is $N_n \equiv N|a_n|^2$. Alternatively, if we had N non-interacting particles in a single experiment, we are expected to find $N|a_n|^2$ of the particles in the same state $|\phi_n\rangle$.

The idea of second quantisation is to replace the amplitudes a_n and a_n^* by operators

$$\sqrt{N}a_n \to \hat{a}_n, \qquad \sqrt{N}a_n^* \to \hat{a}_n^\dagger \tag{A.5}$$

and thus $N_n \to \hat{N}_n = \hat{a}_n^\dagger\hat{a}_n$. Given the canonical form of the equations of motion (A.4), we impose the commutation relations

$$[\hat{a}_n, \hat{a}_m^\dagger] = \delta_{nm}. \tag{A.6}$$

The algebra of the operators is that of a harmonic oscillator with $\hat{a}_n$ ($\hat{a}_n^\dagger$) annihilating (creating) a particle in the state $|\phi_n\rangle$. Furthermore, $\hat{N}_n$ is the number operator of the nth state or mode, such that a Fock state with exactly N_n particles in that mode obeys $\hat{N}_n|N_n\rangle = N_n|N_n\rangle$. The *field operators*

$$\hat{\Psi}(\mathbf{r}) = \sum_n \hat{a}_n\phi_n(\mathbf{r}), \qquad \hat{\Psi}^\dagger(\mathbf{r}) = \sum_n \hat{a}_n^\dagger\phi_n^*(\mathbf{r}), \tag{A.7}$$

respectively, annihilate and create a particle at position $\mathbf{r}$. Their commutation relation follows from (A.6),

$$\left[\hat{\Psi}(\mathbf{r}), \hat{\Psi}^\dagger(\mathbf{r}')\right] = \delta(\mathbf{r} - \mathbf{r}'). \tag{A.8}$$

With these operators we can define the second quantised *particle density* $\hat{n}(\mathbf{r}) = \hat{\Psi}^\dagger(\mathbf{r})\hat{\Psi}(\mathbf{r})$ that gives the number of particles at position $\mathbf{r}$.

The second quantised Hamiltonian is analogously given by

$$\hat{H} = \int d^3\mathbf{r}\,\hat{\Psi}^\dagger(\mathbf{r})\hat{H}_{1\text{st}}\hat{\Psi}(\mathbf{r}), \tag{A.9}$$

where $\hat{H}_{1st}$ denotes the first quantised Hamiltonian. In fact, any first quantised operator $\hat{f}_{1st} = f(\hat{\mathbf{r}}, \hat{\mathbf{p}})$ has its second quantised companion given by the corresponding expression. The Heisenberg equation for the field operator becomes

$$i\hbar \frac{\partial \hat{\Psi}(\mathbf{r}, t)}{\partial t} = \left[\hat{\Psi}(\mathbf{r}, t), \hat{H} \right] = \hat{H}_{1st} \hat{\Psi}(\mathbf{r}, t). \tag{A.10}$$

The interaction between two particles in the first quantisation form is often given in terms of the distance between the two particles, i.e., $V_{1st}(|\mathbf{r}_1 - \mathbf{r}_2|)$. With the coordinates $\mathbf{r}_1$ and $\mathbf{r}_2$ for the two particles, we need to include both particles' field operators $\hat{\Psi}(\mathbf{r}_1)$ and $\hat{\Psi}(\mathbf{r}_2)$. We do not present the details here, but just state the result,

$$\sum_{i<j} V_{1st}(|\mathbf{r}_1 - \mathbf{r}_2|) \rightarrow \frac{1}{2} \int d^2\mathbf{r} d^3\mathbf{r}' \hat{\Psi}^\dagger(\mathbf{r}) \hat{\Psi}^\dagger(\mathbf{r}') V_{1st}(|\mathbf{r}_1 - \mathbf{r}_2|) \hat{\Psi}(\mathbf{r}) \hat{\Psi}(\mathbf{r}').$$

$$\tag{A.11}$$

By including particle–particle interactions on the above form, (A.10) modifies to

$$i\hbar \frac{\partial \hat{\Psi}(\mathbf{r}, t)}{\partial t} = \hat{H}_{1st} \hat{\Psi}(\mathbf{r}, t)$$

$$+ \left[\int d^3\mathbf{r}' \hat{\Psi}^\dagger(\mathbf{r}', t) V_{1st}(|\mathbf{r} - \mathbf{r}'|) \hat{\Psi}(\mathbf{r}', t) \right] \hat{\Psi}(\mathbf{r}, t). \tag{A.12}$$

A few comments are in order. First, while (A.10) reminds much of a Schrödinger equation, it should be remembered that $\hat{\Psi}(\mathbf{r}, t)$ is an operator and the above equation is the Heisenberg equation for this operator. At the mean-field level, we often replace the operators $\hat{\Psi}(\mathbf{r}, t)$ by c-numbers $\Psi(\mathbf{r}, t)$ and the equation is even more reminiscent of the Schrödinger equation. However, $\Psi(\mathbf{r}, t)$ is still not a single particle wave function, but rather the order parameter with the meaning that $|\Psi(\mathbf{r}, t)|^2$ is the particle density. Second, in assigning the commutator (A.6) to the operators, we automatically deal with bosonic particles. For fermionic particles, the method works the same, but we need to change the commutators to anti-commutators, e.g., $\left\{ \hat{a}_n, \hat{a}_m^\dagger \right\} = \hat{a}_n \hat{a}_m^\dagger + \hat{a}_m^\dagger \hat{a}_n = \delta_{nm}$. The fact that different $(n \neq m)$ fermionic operators anti-commute $(\hat{a}_n \hat{a}_m^\dagger = -\hat{a}_m^\dagger \hat{a}_n)$ guarantees that the wave function (A.1) switches sign when two $\mathbf{r}_i$'s are interchanged. If we consider particles with internal degrees of freedom, we need to equip the annihilation and creation operators with a subscript that specifies the internal state. Finally, in cold atom systems, which is the topic of Chap. 5, the atoms are electrically neutral and interact predominantly via s-wave scattering. In this regime, it is legitimate to take the interaction potential to be $V_{1st}(\mathbf{r}_i - \mathbf{r}_j) \propto \delta(\mathbf{r}_i - \mathbf{r}_j)$. Within this approximation, the equations of motion simplify to

$$i\hbar \frac{\partial \hat{\Psi}(\mathbf{r}, t)}{\partial t} = \left[-\frac{\hbar^2}{2m} \nabla^2 + V(\mathbf{r}) + U|\hat{\Psi}(\mathbf{r}, t)|^2 \right] \hat{\Psi}(\mathbf{r}, t), \tag{A.13}$$

where the particles are assumed to move in a potential $V(\mathbf{r})$ and U is the effective interaction strength.

For a periodic potential $V(\mathbf{r}) = V(\mathbf{r}+\mathbf{a})$, we may often expand the field operators in Wannier functions, as discussed in Chap. 4. If we limit the expansion to a single band, we thus have

$$\hat{\Psi}(\mathbf{r}) = \sum_n \hat{a}_n w_{\mathbf{R_n}}(\mathbf{r}). \tag{A.14}$$

If we assume bosons, impose the tight-binding approximation, and use the expansion (A.14), one may derive the Bose-Hubbard many-body Hamiltonian

$$\hat{H}_{\mathrm{BH}} = -J \sum_{\langle ij \rangle} \left(\hat{a}_i^\dagger \hat{a}_j + \mathrm{H.c.} \right) + \frac{U_0}{2} \sum_i \hat{a}_i^\dagger \hat{a}_i^\dagger \hat{a}_i \hat{a}_i. \tag{A.15}$$

The choice of basis $\phi_n(\mathbf{r})$ and truncation of the expansion (such as to consider only a single band) is not always obvious and requires some physical insight.

A.2   Peierls Substitution

The *Peierls substitution* is a practical approximation for handling lattice models in the presence of an external magnetic field characterised by the vector potential $\mathbf{A}(\mathbf{r})$. The method works very well as long as the vector potential can be chosen to be almost uniform on the length scale of the lattice, i.e., over the lattice spacing. Under this assumption, we replace the tunnelling amplitudes of our tight-binding model by

$$t_{ij} \rightarrow t_{ij}\, e^{i\Theta_{ij}}, \qquad \Theta_{ij} = \frac{q}{\hbar} \int_{\mathbf{R_i}}^{\mathbf{R_j}} \mathbf{A}(\mathbf{r}) \cdot d\mathbf{r}, \tag{A.16}$$

where q is the charge of the particle, $\mathbf{R_i}$ is the position of the ith lattice site, and we consider a single band so that the band index ν can be dropped. Upon hopping around a plaquette of a square lattice, say, the accumulated phase would be $\varphi = \Theta_{ij}\Theta_{jk}\Theta_{kl}\Theta_{li}$, as depicted in Fig. A.1, where the hopping occurs between the sites $(i_x, i_y) \rightarrow (i_x + 1, i_y) \rightarrow (i_x + 1, i_y + 1) \rightarrow (i_x, i_y + 1) \rightarrow (i_x, i_y)$. The phase φ is thus the magnetic flux penetrating the plaquette, and thereby a gauge invariant quantity.

We now show how the Peierls substitution comes about when we go back to the derivation of the tight-binding model of Sects. 4.3 and 4.1.4 in the single band case. The expression (4.11) can be inverted so that the Bloch functions are given in terms of the Wannier functions, i.e.,

$$\Psi_{\mathbf{k}}(\mathbf{r}) = \frac{1}{\sqrt{N}} \sum_i e^{i\mathbf{k}\cdot\mathbf{R_i}} w_{\mathbf{R_i}}(\mathbf{r}). \tag{A.17}$$

Fig. A.1 Schematic picture
of the Peierls substitution.
The tunnelling amplitudes t
are multiplied by the phase
factors $e^{i\Theta_{ij}}$ where the phase
Θ_{ij} equals the line integral of
the vector potential $\mathbf{A}(\mathbf{r})$
between the two sites at $\mathbf{R_i}$
and $\mathbf{R_j}$. The total phase
acquired when a particle hops
around a plaquette is thereby
$\varphi = \Theta_{ij}\Theta_{jk}\Theta_{kl}\Theta_{li}$, which
equals the magnetic flux
through the plaquette

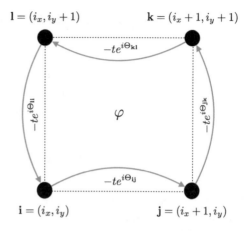

$\mathbf{l} = (i_x, i_y + 1)$ $\mathbf{k} = (i_x + 1, i_y + 1)$

$-te^{i\Theta_{kl}}$

$-te^{i\Theta_{li}}$ φ $-te^{i\Theta_{jk}}$

$-te^{i\Theta_{ij}}$

$\mathbf{i} = (i_x, i_y)$ $\mathbf{j} = (i_x + 1, i_y)$

Thus, the eigenstate of the Hamiltonian with a periodic potential $V(\mathbf{r})$ can be written
on this form. What is the eigenstate of this Hamiltonian when a magnetic field is
present? To answer this, we have to find the eigenstates of the new Hamiltonian

$$\hat{H}' = \frac{(\mathbf{p} - q\mathbf{A}(\mathbf{r}))^2}{2m} + V(\mathbf{r}). \tag{A.18}$$

It turns out that the eigenstates also have a Bloch function structure,

$$\Psi'_{\mathbf{k}}(\mathbf{r}) = \frac{1}{\sqrt{N}} \sum_{\mathbf{i}} e^{i\mathbf{k}\cdot\mathbf{R_i}} w'_{\mathbf{R_i}}(\mathbf{r}) \tag{A.19}$$

with the new 'Wannier functions'

$$w'_{\mathbf{R_i}}(\mathbf{r}) = e^{i\frac{q}{\hbar}\int_{\mathbf{R_i}}^{\mathbf{r}} \mathbf{A}(\mathbf{r}')\cdot d\mathbf{r}'} w_{\mathbf{R_i}}(\mathbf{r}). \tag{A.20}$$

By construction we have

$$\hat{H}' w'_{\mathbf{R_i}}(\mathbf{r}) = \left[\frac{(\mathbf{p} - q\mathbf{A}(\mathbf{r}))^2}{2m} + V(\mathbf{r}) \right] e^{i\frac{q}{\hbar}\int_{\mathbf{R_i}}^{\mathbf{r}} \mathbf{A}(\mathbf{r}')\cdot d\mathbf{r}'} w_{\mathbf{R_i}}(\mathbf{r})$$

$$= e^{i\frac{q}{\hbar}\int_{\mathbf{R_i}}^{\mathbf{r}} \mathbf{A}(\mathbf{r}')\cdot d\mathbf{r}'} \left[\frac{(\mathbf{p} - q\mathbf{A}(\mathbf{r}) + q\mathbf{A}(\mathbf{r}))^2}{2m} + V(\mathbf{r}) \right] w_{\mathbf{R_i}}(\mathbf{r})$$

$$= e^{i\frac{q}{\hbar}\int_{\mathbf{R_i}}^{\mathbf{r}} \mathbf{A}(\mathbf{r}')\cdot d\mathbf{r}'} \hat{H} w_{\mathbf{R_i}}(\mathbf{r}) \tag{A.21}$$

with $\hat{H}$ the 'field-free' Hamiltonian. Multiplying from the left by $\tilde{w}_{\mathbf{R_j}}(\mathbf{r})$, we derive the corresponding tunnelling amplitudes

$$
\begin{aligned}
t'_{ij} &= -\int d^3\mathbf{r}\, w'_{\mathbf{R_i}}(\mathbf{r})\hat{H}'w'_{\mathbf{R_j}}(\mathbf{r}) \\
&= -\int d^3\mathbf{r}\, e^{i\frac{q}{\hbar}\left[\int_{\mathbf{R_i}}^{\mathbf{r}}\mathbf{A}(\mathbf{r}')\cdot d\mathbf{r}' - \int_{\mathbf{R_j}}^{\mathbf{r}}\mathbf{A}(\mathbf{r}')\cdot d\mathbf{r}'\right]}w_{\mathbf{R_i}}(\mathbf{r})\hat{H}w_{\mathbf{R_j}}(\mathbf{r}) \\
&= -e^{i\frac{q}{\hbar}\int_{\mathbf{R_i}}^{\mathbf{R_j}}\mathbf{A}(\mathbf{r}')\cdot d\mathbf{r}'}\int d^3\mathbf{r}\, e^{i\frac{q}{\hbar}\oint\mathbf{A}(\mathbf{r}')\cdot d\mathbf{r}'}w_{\mathbf{R_i}}(\mathbf{r})\hat{H}w_{\mathbf{R_j}}(\mathbf{r}) \approx t_{ij}e^{i\Theta_{ij}}, \quad (A.22)
\end{aligned}
$$

where the contour integral $\oint\mathbf{A}(\mathbf{r}')\cdot d\mathbf{r}'$ is taken along the path $\mathbf{R_i} \to \mathbf{r} \to \mathbf{R_j} \to \mathbf{R_i}$ and is approximated to zero under the assumption that the vector potential can be chosen to be almost uniform on the lattice length scale. Thus, by assuming this, we find that the new tunnelling amplitudes are given by the old ones times a phase factor determined by the vector potential.

A.2.1 Hofstadter Butterfly

By continuing the discussion above, we now consider a square lattice with a homogenous perpendicular magnetic field such that the Hamiltonian, after the Peierls substitution, takes the form

$$
\hat{H}_{\text{Hof}} = -t\sum_{\langle ij \rangle}\left(e^{-i\Theta_{ij}}\hat{a}_i^\dagger\hat{a}_j + h.c.\right), \quad (A.23)
$$

where the phases Θ_{ij} around any single plaquette sum up to φ. Thus, we assume that the magnetic flux per plaquette is φ. For a general flux φ, translational invariance is broken and the band spectrum lost, i.e., the quasi-momenta k_x and k_y and band index ν are no longer good quantum numbers. However, if we write $\varphi = 2\pi\alpha$ it follows that if α is a rational number $\alpha = \frac{p}{q}$, $p, q \in \mathbb{Z}$, the model is still translationally invariant. In particular, the flux through q plaquettes is $2\pi p$, which is equivalent to a zero flux. The new unit cell thus contains q plaquettes. As an example, we consider $q = 4$ in Fig. A.2, and note that there are three possible ways to construct the new unit cells in terms of the old. For instance, in the first, the periodicity is the same in the x direction, while it is $1/4$ in the y-direction. The corresponding first Brillouin zone is thereby changed to $-\pi/a \leq k_x < \pi/a$ and $-\pi/(qa) \leq k_y < \pi/(qa)$. Furthermore, each unit cell contains four sites and the single energy band has split into four bands. More generally, for other values of q we have q bands.

We may limit $\alpha \leq 2\pi$, and we have seen that for any rational α the spectrum of (A.23) comprises a set of bands with varying Brillouin zone sizes. This gives the spectrum a fractal structure when plotted as a function of the flux φ as demonstrated

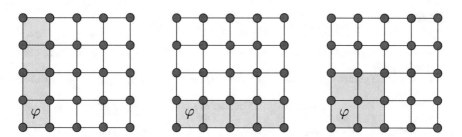

Fig. A.2 The possible unit cells for a square lattice with a uniform plaquette flux $\varphi = 2\pi\alpha$ with $\alpha = 1/4$. For a rational $\alpha = p/q$ ($p, q \in \mathbb{Z}$) the Hamiltonian is translationally invariant with a unit cell of size q times the unit cell with $\varphi = 0$. For $\varphi = 2\pi p/q$ the spectrum splits up in q bands, while for an irrational α the translational invariance is broken and the band structure of the spectrum is lost

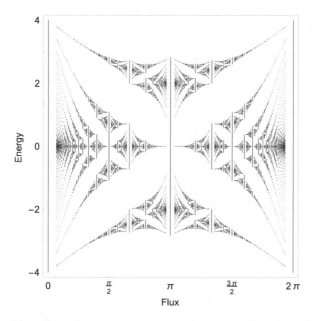

Fig. A.3 The Hofstadter butterfly spectrum of the Hamiltonian (A.23) as a function of the plaquette flux φ. For a general flux φ, translational invariance is broken and the spectrum is discrete, but whenever φ is a rational number times 2π the band structure of the spectrum is reestablished (figure taken from [3])

in Fig. A.3. Due to this characteristic shape, the spectrum has been named the *Hofstadter butterfly* [4].

The continuum limit of the model corresponds to a particle in two dimensions exposed to a perpendicular homogenous magnetic field. This well-known problem is analytically solvable [5], with the equidistant spectrum of infinitely degenerate *Landau levels*. The clustering of eigenenergies in the lower left or right corners of Fig. A.3 are actually remnants of such Landau levels.

References

1. Altland, A., Simons, B.: Condensed Matter Field Theory. Cambridge University Press, Cambridge (2010)
2. Nagaosa, N.: Quantum Field Theory in Condensed Matter Systems. Springer, Berlin/Heidelberg (1999)
3. Hatsuda, Y., Katsura, H., Tachikawa, Y.: Hofstadter's butterfly in quantum geometry. New J. Phys. **18**, 103023 (2016)
4. Hofstadter, D.R.: Energy levels and wave functions of Bloch electrons in rational and irrational magnetic fields. Phys. Rev. B **14**, 2239 (1976)
5. Ballentine, L.: Quantum Mechanics - A Modern Development. World Scientific, Singapore (1976)

Index

© Springer Nature Switzerland AG 2020
J. Larson et al., *Conical Intersections in Physics*, Lecture Notes in Physics 965,
https://doi.org/10.1007/978-3-030-34882-3

Printed in the United States
By Bookmasters